Introduction to Topology

Introduction to Topology

Edited by
Miriam Court

Larsen & Keller
www.larsen-keller.com

Introduction to Topology
Edited by Miriam Court
ISBN: 978-1-63549-704-5 (Hardback)

Larsen & Keller

Published by Larsen and Keller Education,
5 Penn Plaza,
19th Floor,
New York, NY 10001, USA

Cataloging-in-Publication Data

Introduction to topology / edited by Miriam Court.
p. cm.
Includes bibliographical references and index.
ISBN 978-1-63549-704-5
1. Topology. 2. Mathematics. I. Court, Miriam.
QA611 .I58 2018
514--dc23

Table of Contents

Preface

Topology is a sub-field of mathematics. It deals with the study of preserved properties of space under repeated crumpling, stretching and bending. This area includes two major properties namely compactness, and connectedness. The important sub-fields included in topology are geometric topology, general topology, differential topology and algebraic topology. This book unfolds the innovative aspects of topology, which will be crucial for the holistic understanding of the subject matter. The topics included in it are of utmost significance and bound to provide incredible insights to readers. Those in search of information to further their knowledge will be greatly assisted by this textbook.

To facilitate a deeper understanding of the contents of this book a short introduction of every chapter is written below:

Chapter 1- Topology is derived from geometry and set theory. The various sub-fields related to topology are algebraic topology, geometric topology, general topology and differential topology. This is an introductory chapter which will introduce briefly all the significant aspects of topology.

Chapter 2- A group of topological spaces containing a natural topology is known as product topology. Box topology and quotient space are other categories that are discussed here. This chapter elucidates the crucial theories and principles of topology.

Chapter 3- When spaces other than disjointed open subsets are connected together, it is said to be a connected topological space. Spaces can be arc-connected, path-connected and locally connected. Topology is best understood in confluence with the major topics listed in the following chapter.

Chapter 4- Euclidean space includes the two-dimensional Euclidean plane and the three-dimensional Euclidean geometric plane and other spaces. Topological applications of Euclidean space is used to solve problems of Newtonian mechanics. The topics discussed in the chapter are of great importance to broaden the existing knowledge on topology.

Chapter 5- A set of axioms, along with a number of statements, called postulates which are assumed to be "true" are called axiomatic systems. An axiomatic system is said to be complete if every statement or its negative is derivable. The aspects elucidated in this chapter are of vital importance, and provide a better understanding of topology.

I would like to share the credit of this book with my editorial team who worked tirelessly on this book. I owe the completion of this book to the never-ending support of my family, who supported me throughout the project.

Editor

1

Topology and Topological Spaces

Topology is derived from geometry and set theory. The various sub-fields related to topology are algebraic topology, geometric topology, general topology and differential topology. This is an introductory chapter which will introduce briefly all the significant aspects of topology.

Topology

Möbius strips, which have only one surface and one edge, are a kind of object studied in topology.

In mathematics, topology is concerned with the properties of space that are preserved under continuous deformations, such as stretching, crumpling and bending, but not tearing or gluing. This can be studied by considering a collection of subsets, called open sets, that satisfy certain properties, turning the given set into what is known as a topological space. Important topological properties include connectedness and compactness.

Topology developed as a field of study out of geometry and set theory, through analysis of concepts such as space, dimension, and transformation. Such ideas go back to Gottfried Leibniz, who in the 17th century envisioned the *geometria situs* (Greek-Latin for "geometry of place") and *analysis situs* (Greek-Latin for "picking apart of place"). Leonhard Euler's Seven Bridges of Königsberg Problem and Polyhedron Formula are arguably the field's first theorems. The term *topology* was introduced by Johann Benedict Listing in the 19th century, although it was not until the first decades of the 20th century that the idea of a topological space was developed. By the middle of the 20th century, topology had become a major branch of mathematics.

Topology has many subfields:

- General topology, also called point-set topology, establishes the foundational aspects of topology and investigates properties of topological spaces and concepts inherent to topological spaces. It defines the basic notions used in all other branches of topology (including concepts like compactness and connectedness).
- Algebraic topology tries to measure degrees of connectivity using algebraic constructs such as homology and homotopy groups.
- Differential topology is the field dealing with differentiable functions on differentiable manifolds. It is closely related to differential geometry and together they make up the geometric theory of differentiable manifolds.
- Geometric topology primarily studies manifolds and their embeddings (placements) in other manifolds. A particularly active area is low-dimensional topology, which studies manifolds of four or fewer dimensions. This includes knot theory, the study of mathematical knots.

A three-dimensional depiction of a thickened trefoil knot, the simplest non-trivial knot

History

Topology, as a well-defined mathematical discipline, originates in the early part of the twentieth century, but some isolated results can be traced back several centuries. Among these are certain questions in geometry investigated by Leonhard Euler. His 1736 paper on the Seven Bridges of Königsberg is regarded as one of the first practical applications of topology. On 14 November 1750 Euler wrote to a friend that he had realised the importance of the *edges* of a polyhedron. This led to his polyhedron formula, $V - E + F = 2$ (where V, E and F respectively indicate the number of vertices, edges and faces of the polyhedron). Some authorities regard this analysis as the first theorem, signalling the birth of topology.

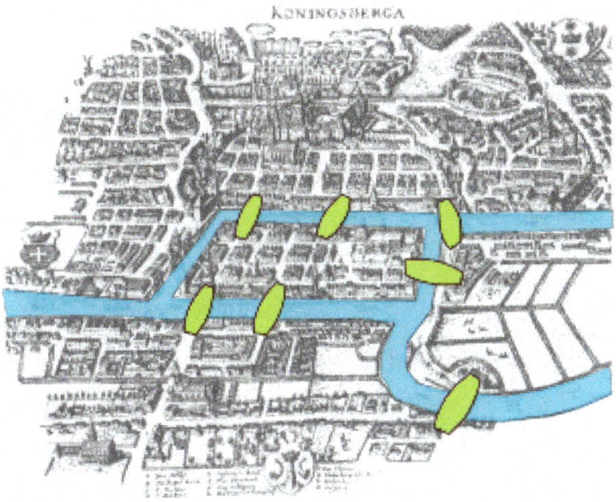

The Seven Bridges of Königsberg was a problem solved by Euler.

Further contributions were made by Augustin-Louis Cauchy, Ludwig Schläfli, Johann Benedict Listing, Bernhard Riemann and Enrico Betti. Listing introduced the term "Topologie" in *Vorstudien zur Topologie*, written in his native German, in 1847, having used the word for ten years in correspondence before its first appearance in print. The English form "topology" was used in 1883 in Listing's obituary in the journal *Nature* to distinguish "...qualitative geometry from the ordinary geometry in which quantitative relations chiefly are treated." The term "topologist" in the sense of a specialist in topology was used in 1905 in the magazine *Spectator*.

Their work was corrected, consolidated and greatly extended by Henri Poincaré. In 1895 he published his ground-breaking paper on *Analysis Situs*, which introduced the concepts now known as homotopy and homology, which are now considered part of algebraic topology.

Topological characteristics of closed 2-manifolds						
Manifold	Euler No. χ	Orientability	Betti numbers			Torsion coefficient (1-dimensional)
			b_0	b_1	b_2	
Sphere	2	Orientable	1	0	1	none
Torus	0	Orientable	1	2	1	none
2-holed torus	−2	Orientable	1	4	1	none
g-holed torus (Genus = g)	2 − 2g	Orientable	1	2g	1	none
Projective plane	1	Non-orientable	1	0	0	2
Klein bottle	0	Non-orientable	1	1	0	2
Sphere with c cross-caps	2 − c	Non-orientable	1	c − 1	0	2
2 Manifold with g holes and c cross-caps (c > 0)	2 − (2g + c)	Non-orientable	1	(2g + c) − 1	0	2

Unifying the work on function spaces of Georg Cantor, Vito Volterra, Cesare Arzelà, Jacques Hadamard, Giulio Ascoli and others, Maurice Fréchet introduced the metric space in 1906. A metric space is now considered a special case of a general topological space, with any given topological space potentially giving rise to many distinct metric spaces. In 1914, Felix Hausdorff coined the term "topological space" and gave the definition for what is now called a Hausdorff space. Currently, a topological space is a slight generalization of Hausdorff spaces, given in 1922 by Kazimierz Kuratowski.

Modern topology depends strongly on the ideas of set theory, developed by Georg Cantor in the later part of the 19th century. In addition to establishing the basic ideas of set theory, Cantor considered point sets in Euclidean space as part of his study of Fourier series.

Introduction

Topology can be formally defined as "the study of qualitative properties of certain objects (called topological spaces) that are invariant under a certain kind of transformation (called a continuous map), especially those properties that are invariant under a certain kind of invertible transformation (called homeomorphism)."

Topology is also used to refer to a structure imposed upon a set X, a structure that essentially 'characterizes' the set X as a topological space by taking proper care of properties such as convergence, connectedness and continuity, upon transformation.

Topological spaces show up naturally in almost every branch of mathematics. This has made topology one of the great unifying ideas of mathematics.

The motivating insight behind topology is that some geometric problems depend not on the exact shape of the objects involved, but rather on the way they are put together. For example, the square and the circle have many properties in common: they are both one dimensional objects (from a topological point of view) and both separate the plane into two parts, the part inside and the part outside.

In one of the first papers in topology, Leonhard Euler demonstrated that it was impossible to find a route through the town of Königsberg (now Kaliningrad) that would cross each of its seven bridges exactly once. This result did not depend on the lengths of the bridges, nor on their distance from one another, but only on connectivity properties: which bridges connect to which islands or riverbanks. This problem in introductory mathematics called *Seven Bridges of Königsberg* led to the branch of mathematics known as graph theory.

Similarly, the hairy ball theorem of algebraic topology says that "one cannot comb the hair flat on a hairy ball without creating a cowlick." This fact is immediately convincing to most people, even though they might not recognize the more formal statement of the theorem, that there is no nonvanishing continuous tangent vector field on the sphere. As with the *Bridges of Königsberg*, the result does not depend on the shape of the sphere; it applies to any kind of smooth blob, as long as it has no holes.

To deal with these problems that do not rely on the exact shape of the objects, one must be clear about just what properties these problems *do* rely on. From this need arises the notion of homeomorphism. The impossibility of crossing each bridge just once applies to any arrangement of bridges homeomorphic to those in Königsberg, and the hairy ball theorem applies to any space homeomorphic to a sphere.

Intuitively, two spaces are homeomorphic if one can be deformed into the other without cutting or gluing. A traditional joke is that a topologist cannot distinguish a coffee mug from a doughnut, since a sufficiently pliable doughnut could be reshaped to a coffee cup by creating a dimple and progressively enlarging it, while shrinking the hole into a handle.

Homeomorphism can be considered the most basic *topological equivalence*. Another is homotopy equivalence. This is harder to describe without getting technical, but the essential notion is that two objects are homotopy equivalent if they both result from "squishing" some larger object.

Equivalence classes of the English (i.e., Latin) alphabet (sans-serif)	
Homeomorphism	Homotopy equivalence
{A,R} {B} {C,G,I,J,L,M,N,S,U,V,W,Z} {D,O} {E,F,T,Y} {H,K} {P,Q} {X}	{A,R,D,O,P,Q} {B} {C,E,F,G,H,I,J,K,L,M,N,S,T,U,V,W,X,Y,Z}

An introductory exercise is to classify the uppercase letters of the English alphabet according to homeomorphism and homotopy equivalence. The result depends partially on the font used. The figures use the sans-serif Myriad font. Homotopy equivalence is a rougher relationship than homeomorphism; a homotopy equivalence class can contain several homeomorphism classes. The simple case of homotopy equivalence described above can be used here to show two letters are homotopy equivalent. For example, O fits inside P and the tail of the P can be squished to the "hole" part.

Homeomorphism classes are:

- no holes,
- no holes three tails,
- no holes four tails,
- one hole no tail,
- one hole one tail,
- one hole two tails,

- two holes no tail, and
- a bar with four tails.

Homotopy classes are larger, because the tails can be squished down to a point. They are:

- one hole,
- two holes, and
- no holes.

To classify the letters correctly, we must show that two letters in the same class are equivalent and two letters in different classes are not equivalent. In the case of homeomorphism, this can be done by selecting points and showing their removal disconnects the letters differently. For example, X and Y are not homeomorphic because removing the center point of the X leaves four pieces; whatever point in Y corresponds to this point, its removal can leave at most three pieces. The case of homotopy equivalence is harder and requires a more elaborate argument showing an algebraic invariant, such as the fundamental group, is different on the supposedly differing classes.

Letter topology has practical relevance in stencil typography. For instance, Braggadocio font stencils are made of one connected piece of material.

Concepts

Topologies on Sets

The term topology also refers to a specific mathematical idea central to the area of mathematics called topology. Informally, a topology tells how elements of a set relate spatially to each other. The same set can have different topologies. For instance, the real line, the complex plane (which is a 1-dimensional complex vector space), and the Cantor set can be thought of as the same set with different topologies.

Formally, let X be a set and let τ be a family of subsets of X. Then τ is called a *topology on X* if:

1. Both the empty set and X are elements of τ.
2. Any union of elements of τ is an element of τ.
3. Any intersection of finitely many elements of τ is an element of τ.

If τ is a topology on X, then the pair (X, τ) is called a *topological space*. The notation X_τ may be used to denote a set X endowed with the particular topology τ.

The members of τ are called *open sets* in X. A subset of X is said to be closed if its complement is in τ (i.e., its complement is open). A subset of X may be open, closed, both (clopen set), or neither. The empty set and X itself are always both closed and open. An open set containing a point x is called a 'neighborhood' of x.

A set with a topology is called a topological space.

Continuous Functions and Homeomorphisms

A function or map from one topological space to another is called *continuous* if the inverse image of any open set is open. If the function maps the real numbers to the real numbers (both spaces with the Standard Topology), then this definition of continuous is equivalent to the definition of continuous in calculus. If a continuous function is one-to-one and onto, and if the inverse of the function is also continuous, then the function is called a homeomorphism and the domain of the function is said to be homeomorphic to the range. Another way of saying this is that the function has a natural extension to the topology. If two spaces are homeomorphic, they have identical topological properties, and are considered topologically the same. The cube and the sphere are homeomorphic, as are the coffee cup and the doughnut. But the circle is not homeomorphic to the doughnut.

Manifolds

While topological spaces can be extremely varied and exotic, many areas of topology focus on the more familiar class of spaces known as manifolds. A manifold is a topological space that resembles Euclidean space near each point. More precisely, each point of an n-dimensional manifold has a neighbourhood that is homeomorphic to the Euclidean space of dimension n. Lines and circles, but not figure eights, are one-dimensional manifolds. Two-dimensional manifolds are also called surfaces. Examples include the plane, the sphere, and the torus, which can all be realized without self-intersection in three dimensions, but also the Klein bottle and real projective plane, which cannot.

Topics

General Topology

General topology is the branch of topology dealing with the basic set-theoretic definitions and constructions used in topology. It is the foundation of most other branches of topology, including differential topology, geometric topology, and algebraic topology. Another name for general topology is point-set topology.

The fundamental concepts in point-set topology are *continuity*, *compactness*, and *connectedness*. Intuitively, continuous functions take nearby points to nearby points. Compact sets are those that can be covered by finitely many sets of arbitrarily small size. Connected sets are sets that cannot be divided into two pieces that are far apart. The words *nearby*, *arbitrarily small*, and *far apart* can all be made precise by using open sets. If we change the definition of *open set*, we change what continuous functions, compact sets, and connected sets are. Each choice of definition for *open set* is called a *topology*. A set with a topology is called a *topological space*.

Metric spaces are an important class of topological spaces where distances can be assigned a number called a *metric*. Having a metric simplifies many proofs, and many of the most common topological spaces are metric spaces.

Algebraic Topology

Algebraic topology is a branch of mathematics that uses tools from abstract algebra to study topological spaces. The basic goal is to find algebraic invariants that classify topological spaces up to homeomorphism, though usually most classify up to homotopy equivalence.

The most important of these invariants are homotopy groups, homology, and cohomology.

Although algebraic topology primarily uses algebra to study topological problems, using topology to solve algebraic problems is sometimes also possible. Algebraic topology, for example, allows for a convenient proof that any subgroup of a free group is again a free group.

Differential Topology

Differential topology is the field dealing with differentiable functions on differentiable manifolds. It is closely related to differential geometry and together they make up the geometric theory of differentiable manifolds.

More specifically, differential topology considers the properties and structures that require only a smooth structure on a manifold to be defined. Smooth manifolds are 'softer' than manifolds with extra geometric structures, which can act as obstructions to certain types of equivalences and deformations that exist in differential topology. For instance, volume and Riemannian curvature are invariants that can distinguish different geometric structures on the same smooth manifold—that is, one can smoothly "flatten out" certain manifolds, but it might require distorting the space and affecting the curvature or volume.

Geometric Topology

Geometric topology is a branch of topology that primarily focuses on low-dimensional manifolds (i.e. dimensions 2,3 and 4) and their interaction with geometry, but it also includes some higher-dimensional topology. Some examples of topics in geometric topology are orientability, handle decompositions, local flatness, crumpling and the planar and higher-dimensional Schönflies theorem.

In high-dimensional topology, characteristic classes are a basic invariant, and surgery theory is a key theory.

Low-dimensional topology is strongly geometric, as reflected in the uniformization theorem in 2 dimensions – every surface admits a constant curvature metric; geometrically, it has one of 3 possible geometries: positive curvature/spherical, zero curvature/flat, negative curvature/hyperbolic – and the geometrization conjecture (now theorem) in 3 dimensions – every 3-manifold can be cut into pieces, each of which has one of eight possible geometries.

2-dimensional topology can be studied as complex geometry in one variable (Riemann surfaces are complex curves) – by the uniformization theorem every conformal class of metrics is equivalent to a unique complex one, and 4-dimensional topology can be studied from the point of view of complex geometry in two variables (complex surfaces), though not every 4-manifold admits a complex structure.

Generalizations

Occasionally, one needs to use the tools of topology but a "set of points" is not available. In pointless topology one considers instead the lattice of open sets as the basic notion of the theory, while Grothendieck topologies are structures defined on arbitrary categories that allow the definition of sheaves on those categories, and with that the definition of general cohomology theories.

Applications

Biology

Knot theory, a branch of topology, is used in biology to study the effects of certain enzymes on DNA. These enzymes cut, twist, and reconnect the DNA, causing knotting with observable effects such as slower electrophoresis. Topology is also used in evolutionary biology to represent the relationship between phenotype and genotype. Phenotypic forms that appear quite different can be separated by only a few mutations depending on how genetic changes map to phenotypic changes during development.

Computer Science

Topological data analysis uses techniques from algebraic topology to determine the large scale structure of a set (for instance, determining if a cloud of points is spherical or toroidal). The main method used by topological data analysis is:

1. Replace a set of data points with a family of simplicial complexes, indexed by a proximity parameter.
2. Analyse these topological complexes via algebraic topology – specifically, via the theory of persistent homology.
3. Encode the persistent homology of a data set in the form of a parameterized version of a Betti number, which is called a barcode.

Physics

In physics, topology is used in several areas such as condensed matter physics, quantum field theory and physical cosmology.

The topological dependence of mechanical properties in solids is of interest in disciplines of mechanical engineering and materials science. Electrical and mechanical properties depend on the arrangement and network structures of molecules and elementary units in materials. The compressive strength of crumpled topologies is studied in attempts to understand the high strength to weight of such structures that are mostly empty space. Topology is of further significance in Contact mechanics where the dependence of stiffness and friction on the dimensionality of surface structures is the subject of interest with applications in multi-body physics.

A topological quantum field theory (or topological field theory or TQFT) is a quantum field theory that computes topological invariants.

Although TQFTs were invented by physicists, they are also of mathematical interest, being related to, among other things, knot theory and the theory of four-manifolds in algebraic topology, and to the theory of moduli spaces in algebraic geometry. Donaldson, Jones, Witten, and Kontsevich have all won Fields Medals for work related to topological field theory.

The topological classification of Calabi-Yau manifolds has important implications in string theory, as different manifolds can sustain different kinds of strings.

In cosmology, topology can be used to describe the overall shape of the universe. This area is known as spacetime topology.

Robotics

The various possible positions of a robot can be described by a manifold called configuration space. In the area of motion planning, one finds paths between two points in configuration space. These paths represent a motion of the robot's joints and other parts into the desired location and pose.

Topological Space

In topology and related branches of mathematics, a topological space may be defined as a set of points, along with a set of neighbourhoods for each point, satisfying a set of axioms relating points and neighbourhoods. The definition of a topological space relies only upon set theory and is the most general notion of a mathematical space that allows for the definition of concepts such as continuity, connectedness, and convergence.

Other spaces, such as manifolds and metric spaces, are specializations of topological spaces with extra structures or constraints. Being so general, topological spaces are a central unifying notion and appear in virtually every branch of modern mathematics. The branch of mathematics that studies topological spaces in their own right is called point-set topology or general topology.

Definition

The utility of the notion of a topology is shown by the fact that there are several equivalent definitions of this structure. Thus one chooses the axiomatisation suited for the application. The most commonly used, and the most elegant, is that in terms of *open sets*, but the most intuitive is that in terms of *neighbourhoods* and so this is given first. Note: A variety of other axiomatisations of topological spaces are listed in the Exercises of the book by Vaidyanathaswamy.

Definition via Neighbourhoods

This axiomatization is due to Felix Hausdorff. Let X be a set; the elements of X are usually called *points*, though they can be any mathematical object. We allow X to be empty. Let N be a function assigning to each x (point) in X a non-empty collection N(x) of subsets of X. The elements of N(x) will be called *neighbourhoods* of x with respect to N (or, simply, *neighbourhoods of* x). The function N is called a neighbourhood topology if the axioms below are satisfied; and then X with N is called a topological space.

1. If N is a neighbourhood of x (i.e., $N \in$ N(x)), then x $\in N$. In other words, each point belongs to every one of its neighbourhoods.

2. If N is a subset of X and includes a neighbourhood of x, then N is a neighbourhood of x. I.e., every superset of a neighbourhood of a point x in X is again a neighbourhood of x.

3. The intersection of two neighbourhoods of x is a neighbourhood of x.

4. Any neighbourhood N of x includes a neighbourhood M of x such that N is a neighbourhood of each point of M.

The first three axioms for neighbourhoods have a clear meaning. The fourth axiom has a very important use in the structure of the theory, that of linking together the neighbourhoods of different points of X.

A standard example of such a system of neighbourhoods is for the real line R, where a subset N of R is defined to be a *neighbourhood* of a real number x if it includes an open interval containing x.

Given such a structure, we can define a subset U of X to be open if U is a neighbourhood of all points in U. It is a remarkable fact that the open sets then satisfy the elegant axioms given below, and that, given these axioms, we can recover the neighbourhoods satisfying the above axioms by defining N to be a neighbourhood of x if N includes an open set U such that $x \in U$.

Definition via Open Sets

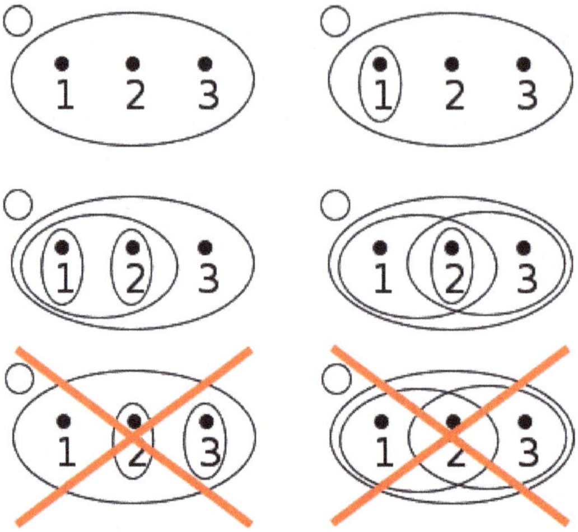

Four examples and two non-examples of topologies on the three-point set {1,2,3}. The bottom-left example is not a topology because the union of {2} and {3} [i.e. {2,3}] is missing; the bottom-right example is not a topology because the intersection of {1,2} and {2,3} [i.e. {2}], is missing.

A *topological space* is an ordered pair (X, τ), where X is a set and τ is a collection of subsets of X, satisfying the following axioms:

1. The empty set and X itself belong to τ.
2. Any (finite or infinite) union of members of τ still belongs to τ.
3. The intersection of any finite number of members of τ still belongs to τ.

The elements of τ are called open sets and the collection τ is called a topology on X.

Examples

1. Given $X = \{1, 2, 3, 4\}$, the collection $\tau = \{\{\}, \{1, 2, 3, 4\}\}$ of only the two subsets of X required by the axioms forms a topology of X, the trivial topology (indiscrete topology).
2. Given $X = \{1, 2, 3, 4\}$, the collection $\tau = \{\{\}, \{2\}, \{1, 2\}, \{2, 3\}, \{1, 2, 3\}, \{1, 2, 3, 4\}\}$ of six subsets of X forms another topology of X.
3. Given $X = \{1, 2, 3, 4\}$ and the collection $\tau = P(X)$ (the power set of X), (X, τ) is a topological space. τ is called the discrete topology.

4. Given $X = Z$, the set of integers, the collection τ of all finite subsets of the integers plus Z itself is *not* a topology, because (for example) the union of all finite sets not containing zero is infinite but is not all of Z, and so is not in τ.

Definition via Closed Sets

Using de Morgan's laws, the above axioms defining open sets become axioms defining closed sets:

1. The empty set and X are closed.
2. The intersection of any collection of closed sets is also closed.
3. The union of any finite number of closed sets is also closed.

Using these axioms, another way to define a topological space is as a set X together with a collection τ of closed subsets of X. Thus the sets in the topology τ are the closed sets, and their complements in X are the open sets.

Other Definitions

There are many other equivalent ways to define a topological space: in other words, the concepts of neighbourhood or of open or closed sets can be reconstructed from other starting points and satisfy the correct axioms.

Another way to define a topological space is by using the Kuratowski closure axioms, which define the closed sets as the fixed points of an operator on the power set of X.

A net is a generalisation of the concept of sequence. A topology is completely determined if for every net in X the set of its accumulation points is specified.

Comparison of Topologies

A variety of topologies can be placed on a set to form a topological space. When every set in a topology τ_1 is also in a topology τ_2 and τ_1 is a subset of τ_2, we say that τ_2 is *finer* than τ_1, and τ_1 is *coarser* than τ_2. A proof that relies only on the existence of certain open sets will also hold for any finer topology, and similarly a proof that relies only on certain sets not being open applies to any coarser topology. The terms *larger* and *smaller* are sometimes used in place of finer and coarser, respectively. The terms *stronger* and *weaker* are also used in the literature, but with little agreement on the meaning, so one should always be sure of an author's convention when reading.

The collection of all topologies on a given fixed set X forms a complete lattice: if $F = \{\tau_\alpha \mid \alpha \text{ in } A\}$ is a collection of topologies on X, then the meet of F is the intersection of F, and the join of F is the meet of the collection of all topologies on X that contain every member of F.

Continuous Functions

A function $f : X \to Y$ between topological spaces is called continuous if for every x $\in X$ and every neighbourhood N of f(x) there is a neighbourhood M of x such that $f(M) \subseteq N$. This relates easily to the usual definition in analysis. Equivalently, f is continuous if the inverse image of every open set is open. This is an attempt to capture the intuition that there are no "jumps" or "separations" in the function. A homeomorphism is a bijection that is continuous and whose inverse is also continuous. Two spaces are called *homeomorphic* if there exists a homeomorphism between them. From the standpoint of topology, homeomorphic spaces are essentially identical.

In category theory, Top, the category of topological spaces with topological spaces as objects and continuous functions as morphisms is one of the fundamental categories in category theory. The attempt to classify the objects of this category (up to homeomorphism) by invariants has motivated areas of research, such as homotopy theory, homology theory, and K-theory etc.

Examples of Topological Spaces

A given set may have many different topologies. If a set is given a different topology, it is viewed as a different topological space. Any set can be given the discrete topology in which every subset is open. The only convergent sequences or nets in this topology are those that are eventually constant. Also, any set can be given the trivial topology (also called the indiscrete topology), in which only the empty set and the whole space are open. Every sequence and net in this topology converges to every point of the space. This example shows that in general topological spaces, limits of sequences need not be unique. However, often topological spaces must be Hausdorff spaces where limit points are unique.

Metric Spaces

Metric spaces embody a metric, a precise notion of distance between points.

Every metric space can be given a metric topology, in which the basic open sets are open balls defined by the metric. This is the standard topology on any normed vector space. On a finite-dimensional vector space this topology is the same for all norms.

There are many ways of defining a topology on R, the set of real numbers. The standard topology on R is generated by the open intervals. The set of all open intervals forms a base or basis for the topology, meaning that every open set is a union of some collection of sets from the base. In particular, this means that a set is open if there exists an open interval of non zero radius about every point in the set. More generally, the Euclidean spaces R^n can be given a topology. In the usual topology on R^n the basic open sets are the open balls. Similarly, C, the set of complex numbers, and C^n have a standard topology in which the basic open sets are open balls.

Topological Constructions

Every subset of a topological space can be given the subspace topology in which the open sets are the intersections of the open sets of the larger space with the subset. For any indexed family of topological spaces, the product can be given the product topology, which is generated by the inverse images of open sets of the factors under the projection mappings. For example, in finite products, a basis for the product topology consists of all products of open sets. For infinite products, there is the additional requirement that in a basic open set, all but finitely many of its projections are the entire space.

A quotient space is defined as follows: if X is a topological space and Y is a set, and if f: $X \rightarrow Y$ is a surjective function, then the quotient topology on Y is the collection of subsets of Y that have open inverse images under f. In other words, the quotient topology is the finest topology on Y for which f is continuous. A common example of a quotient topology is when an equivalence relation is defined on the topological space X. The map f is then the natural projection onto the set of equivalence classes.

The Vietoris topology on the set of all non-empty subsets of a topological space X, named for Leopold Vietoris, is generated by the following basis: for every n-tuple U_1, ..., U_n of open sets in X, we construct a basis set consisting of all subsets of the union of the U_i that have non-empty intersections with each U_i.

The Fell topology on the set of all non-empty closed subsets of a locally compact Polish space X is a variant of the Vietoris topology. It is generated by the following basis: for every n-tuple U_1, ..., U_n of open sets in X and for every compact set K, the set of all subsets of X that are disjoint from K and have nonempty intersections with each U_i is a member of the basis.

Classification of Topological Spaces

Topological spaces can be broadly classified, up to homeomorphism, by their topological properties. A topological property is a property of spaces that is invariant under homeomorphisms. To prove that two spaces are not homeomorphic it is sufficient to find a topological property not shared by them. Examples of such properties include connectedness, compactness, and various separation axioms.

Topological Spaces with Algebraic Structure

For any algebraic objects we can introduce the discrete topology, under which the algebraic operations are continuous functions. For any such structure that is not finite, we often have a natural topology compatible with the algebraic operations, in the sense that the algebraic operations are still continuous. This leads to concepts such as topological groups, topological vector spaces, topological rings and local fields.

Topological Spaces with Order Structure

- Spectral. A space is spectral if and only if it is the prime spectrum of a ring (Hochster theorem).
- Specialization preorder. In a space the specialization (or canonical) preorder is defined by $x \leq y$ if and only if $cl\{x\} \subseteq cl\{y\}$.

Basic Concepts

We start with the assumption that we intuitively understand what is meant by a set. For us, set is a collection of well defined objects. We have a set X and let $\mathcal{J}$ be a collection of subsets of X satisfying:

- $(T1)\ \phi \in \mathcal{J},\ X \in \mathcal{J}$, where ϕ is the empty set (or say null set).
- (T2) Suppose we have an arbitrary nonempty set J and to each $\alpha \in J$ we have a subset A_α of X such that $A_\alpha \in \mathcal{J}$, then our $\mathcal{J}$ has the property that $\bigcup_{\alpha \in J} A_\alpha \in \mathcal{J}$, where $\bigcup_{\alpha \in J} A_\alpha = \{x \in X : x \in A_\alpha \text{ for at least one } \alpha \in J\}$.
- (T3) If A_1, A_2 are in $\mathcal{J}$ then $A_1 \cap A_2$ is also in $\mathcal{J}$ (that is $A_1, A_2 \in \mathcal{J}$ implies $A_1 \cap A_2 \in \mathcal{J}$).

In such a case, the given collection $\mathcal{J}$ is called a topology on X and the pair $(X, \mathcal{J})$ is called a topological space.

Remark: If A is a member of $\mathcal{J}$ and $x \in A$ then we say that A is a neighbourhood (also known as open neighbourhood) of x: That is for each x in X; $\mathcal{J}$ contains the collection $\mathcal{N}_x = \{U \in \mathcal{J} : x \in U\}$ of all open neighbourhoods of x.

Suppose we are given a set X. Now our aim is to find collections $\mathcal{B}$ and $\mathcal{J}$ of subsets of X satisfying:

(i) $\mathcal{B} \subseteq \mathcal{J}$, (ii) $\mathcal{J}$ satisfies (T1), (T2), (T3), and (iii) $\mathcal{J} = \{U \subseteq X : x \in U$ implies there exists $B \in \mathcal{B}$ such that $x \in B \subseteq U\}$.

In such a case, $\mathcal{J}$ is said to be a topology on X generated by the collection $\mathcal{B}$ and $\mathcal{B}$ is said to be a basis for the topology $\mathcal{J}$. Each member of $\mathcal{J}$ is called an open subset of X and each member of $\mathcal{B}$ is called an essential neighbourhood or a basic open set in X. Since $X \in \mathcal{J}$, by (iii) for each $x \in X$ there exists $B \in \mathcal{B}$ such that $x \in B$. Also note that if $B_1, B_2 \in \mathcal{B}$ then $B_1 \cap B_2 \in \mathcal{J}$. Hence for any $x \in B_1 \cap B_2$ there exists $B_3 \in \mathcal{B}$ such that $x \in B_3 \subseteq B_1 \cap B_2$. Therefore $\mathcal{B}$ satisfies the following:

- (B1) For every $x \in X$ there exists $B \in \mathcal{B}$ such that $x \in B$.
- (B2) $B_1, B_2 \in \mathcal{B}$ and $x \in B_1 \cap B_2$ implies there exists $B_3 \in \mathcal{B}$ such that $x \in B_3 \subseteq B_1 \cap B_2$.

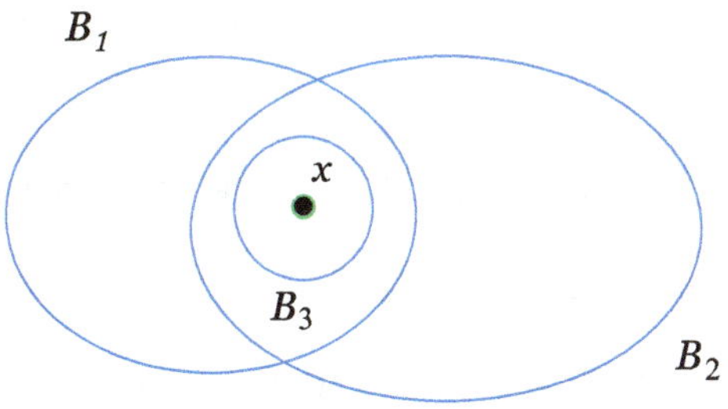

Suppose a collection $\mathscr{B}$ of subsets of a given set X satisfies the conditions (B1), (B2). Then using (iii) we can define $\mathcal{J}$ and such a collection $\mathcal{J}$ satisfies (T1), (T2), and (T3).

Let us prove the following theorem:

Theorem 1. Suppose a collection $\mathscr{B}$ of subsets of a given set X satisfies:

(B1) For every $x \in X$ there exists $B \in \mathscr{B}$ such that $x \in B$.

(B2) $B_1, B_2 \in \mathscr{B}$ and $x \in B_1 \cap B_2$ implies there exists $B_3 \in \mathscr{B}$ such that $x \in B_3 \subseteq B_1 \cap B_2$.

Then the collection $\mathcal{J}$ defined as $\mathcal{J} = \{U \subseteq X : x \in U$ implies there exists $B \in \mathscr{B}$ such that $x \in B \subseteq U\}$ is a topology on X.

Proof: To prove (T1): From (B1), $x \in X$ implies there exists $B \in \mathscr{B}$ such that $x \in B \subseteq X$. Hence by the definition of $\mathcal{J}$, $X \in \mathcal{J}$. Now we will have to prove that the null set $\phi \in \mathcal{J}$. How to prove? Our statement namely $x \in U$ implies there exists $B \in \mathscr{B}$ such that $x \in B \subseteq U$ is a conditional statement. That is, we have statements say p and q. Now consider the truth table

p	q	p $\Rightarrow$ q
T	T	T
T	F	F
F	T	T
F	F	T

The so-called null set ϕ (or empty set) is a subset of X. Whether ϕ satisfies the stated property? What is the stated property with respect to our set ϕ? If $x \in \phi$ then there exists $B \in \mathscr{B}$ such that $x \in B \subseteq \phi$, where are we in the truth table? Whether there is $x \in \phi$? The answer is no. So our statement $x \in \phi$ is false. In such a case whether q is true or false it does not matter and p $\Rightarrow$ q is true. So the conclusion is that the null set ϕ has the stated property, therefore by the definition of $\mathcal{J}$, $\phi \in \mathcal{J}$.

To prove (T2): Suppose J is a nonempty set and for each $\alpha \in J$, $A_\alpha \in \mathcal{J}$. Now we will have to prove that $\bigcup_{\alpha \in J} A_\alpha \in \mathcal{J}$.

If $\bigcup_{\alpha \in J} A_\alpha = \phi$, then $\phi \in \mathcal{J}$ (follows from (T1)). So let us assume that $\bigcup_{\alpha \in J} A_\alpha \neq \phi$. Let $x \in A$, then there exists $\alpha_0 \in J$ such that $x \in A_{\alpha_0}$. Now $x \in A_{\alpha_0}$ and $A_{\alpha_0} \in \mathcal{J}$ therefore by the definition of $\mathcal{J}$ there exists $B \in \mathscr{B}$ such that $x \in B \subseteq A_{\alpha_0}$. But $A_{\alpha_0} \subseteq \bigcup_{\alpha \in J} A_\alpha$. Hence $x \in \bigcup_{\alpha \in J} A_\alpha$ implies there exists $B \in \mathscr{B}$ such that $x \in B \subseteq \bigcup_{\alpha \in J} A_\alpha$ (since $x \in B \subseteq A_{\alpha_0}$).

Therefore by the definition of $\mathcal{J}$, $\bigcup_{\alpha \in J} A_\alpha \in \mathcal{J}$. That is, if J is a nonempty set and $A_\alpha \in \mathcal{J}$ for all $\alpha \in J$ then $\bigcup_{\alpha \in J} A_\alpha \in \mathcal{J}$.

To prove (T3): Let $A_1, A_2 \in \mathcal{J}$. Again if $A_1 \cap A_2 = \phi$ then by (T1), $\phi \in \mathcal{J}$ and hence $A_1 \cap A_2 \in \mathcal{J}$.

Suppose $A_1 \cap A_2 \neq \phi$. Now let $x \in A_1 \cap A_2$ then $x \in A_1$ and $x \in A_2$. Now $x \in A_1$, $A_1 \in \mathcal{J}$ implies there exists $B_1 \in \mathscr{B}$ so that $x \in B_1 \subseteq A_1$. Also $x \in A_2$, $A_2 \in \mathcal{J}$ implies there exists $B_2 \in \mathscr{B}$ such that $x \in B_2 \subseteq A_2$. Now $B_1, B_2 \in \mathscr{B}$ are such that $x \in B_1 \cap B_2$. Hence by (B2) there exists $B_3 \in \mathscr{B}$ such that $x \in B_3 \subseteq B_1 \cap B_2$. But $B_1 \cap B_2 \subseteq A_1 \cap A_2$. Hence $B_3 \in \mathscr{B}$ is such that $x \in B_3 \subseteq A_1 \cap A_2$.

That is $x \in A_1 \cap A_2$ implies there exists $B_3 \in \mathscr{B}$ such that $x \in B_3 \subseteq A_1 \cap A_2$ implies $A_1 \cap A_2 \in \mathcal{J}$ (by the definition of $\mathcal{J}$). Now $\mathcal{J}$ satisfies (T1), (T2), (T3) and therefore $\mathcal{J}$ is a topology on X.

Remark: The topology $\mathcal{J}$ defined as in theorem 1 is called the topology generated by $\mathscr{B}$. If we want to define a topology on a set X then we search for a collection $\mathscr{B}$ of subsets of X satisfying (B1), (B2) and once we know such a collection $\mathscr{B}$ then we know how to get the topology generated by $\mathscr{B}$. Such a collection $\mathscr{B}$ is called a basis for a topology on X and the topology generated by $\mathscr{B}$ is normally denoted by $\mathcal{J}_{\mathscr{B}}$.

Definition: If $\mathscr{B}$ is a collection of subsets of a given set X satisfying (B1), (B2) then $\mathscr{B}$ is called a basis for a topology on X.

The Metric Topology

Let X be a nonempty set and $(x, y) \in X \times X$. With each $(x, y) \in X \times X$ we associate a non-negative real number which we denote by $d(x, y)$. We want to identify $d(x, y)$ as the distance between the elements x, y in X. So it is natural to expect that

- (M1) $d(x, y) = 0$ if and only if $x = y$;

- (M2) $d(x,y)=d(y,x)$ for all $x,\ y \in X$;
- (M3) $d(x,y) \leq d(x,z)+d(z,y)$ for all $x,\ y,\ z \in X$.

It is to be noted that to each element (x, y) in X × X we associate a unique element $d(x, y)$ in $\mathbb{R}^+ = [0, \infty)$. That is $d(x, y)$ is the image of $(x, y) \in X \times X$. Hence d is a function from X × X into $\mathbb{R}^+$ i.e. $d : X \times X \rightarrow \mathbb{R}^+$.

If X is a nonempty set and $d : X \times X \rightarrow \mathbb{R}^+$ is a function satisfying the above conditions (M1), (M2), (M3) then we say that d is a metric on X. In such a case, the pair (X, d) is called a metric space.

Let us fix $x \in X$. Now we want to collect all those elements of the space X which are not far away from x and such a set is known as a neighbourhood of x. Well, what do you mean by "not far away from x"? The term "not far away" is a relative term. So we fix an r > 0 (in some sense radius of our neighbourhood) and then take an element, say y from X. If the distance between x and y is strictly less than r, that is $d(x, y) < r$, then we say that y is in our neighbourhood of x. Let us define $B(x, r) = \{y \in X : d(x, y) < r\}$, and call this set as one of our neighbourhoods of x. If we change r, we get different neighbourhoods of x and B (x, r) is also known as the ball centered at x and radius r. When X=$\mathbb{R}^3$ and d, the distance function, is the usual Euclidean distance, i.e. for any $x = (x_1, x_2, x_3),\ y = (y_1, y_2, y_3) \in \mathbb{R}^3$ $d(x, y) = \sqrt{(x_1 - y_1)^2 + (x_2 - y_2)^2 + (x_3 - y_3)^2}$, then $B(x, r)$ is the usual Euclidean ball centered at x and radius r > 0.

Remark: One can have different metrics on $\mathbb{R}^3 (or\ \mathbb{R}^n,\ n \geq 1)$ and for $x = (x_1, x_2, x_3) \in \mathbb{R}^3,\ r > 0,\ B(x, r)$ may be a cube or a solid sphere or an ellipsoid (excluding the points on the boundary) or a singleton $\{x\}$ or the whole space $\mathbb{R}^3$ under suitable metrics. Now consider a subset A of X. Suppose A has the property: if $x \in A$ then there exists at least one neighbourhood of x say $B(x,r)$ which is contained in our set A. That is, $x \in A$ implies there exists r > 0 such that $B(x, r) \subseteq A$ (such a r > 0 depends on $x \in A$. i.e. same r may not work for every $x \in A$).

Note. Our statement namely $x \in A$ implies there exists r > 0 such that $B(x, r) \subseteq A$ is a conditional statement. The so-called empty set (or null set) ϕ is a subset of our space X. Whether empty set ϕ has the stated property? What is the stated property? Well, following the same argument as given in theorem 1 we see that ϕ has the stated property. Now it is easy (if not obvious) to prove:

- X has the stated property.
- A, B ⊆ X such that A, B have the stated property then A ∩ B has the stated property.

- Consider a nonempty set J. Suppose for each $\alpha \in J$, $A_\alpha \subseteq X$ and A_α has the stated property, then $\bigcup_{\alpha \in J} A_\alpha$ has the stated property. That is the collection J_d defined as $J_d = \{A \subseteq X : x \in A$ implies there exists r > o such that $B(x, r) \subseteq A\}$ is a topology on X, known as the topology induced by the given metric d.

In this sense we say that every metric space (X, d) is a topological space.

Theorem 2. In a metric space (X, d) for each $x \in X$, $r > 0$, $B(x, r)$ is an open subset of (X, J_d).

Proof: Let $y \in B(x, r)$. Then $d(x, y) < r$. Let $s = r - d(x, y)$. If $z \in B(y, s)$, then $d(y, z) < s = r - d(x, y)$. So $d(x,y) + d(y,z) < r$. By the triangle inequality, $d(x,z) < r$. That is $z \in B(x, r)$. Thus $B(y, s) \subset B(x, r)$. Hence $B(x, r)$ is open.

It is interesting to note that $\mathscr{B} = \{B(x, r) : x \in X, r > 0\}$ is a basis for a topology on X and it is clear from the definition of J_d that the topology $J_{\mathscr{B}}$ generated by $\mathscr{B}$ is same as J_d.

Now let us give some important examples of topological spaces.

Let X be a set and let $J_t = \{\phi, X\}$, $J_D = \{A : A \text{ is subset of } X\}$, $J_f = \{A : X \setminus A = A^c$ is a finite subset of X or $A^c = X\}$, $J_c = \{A : X \setminus A$ is a countable subset of A or $A^c = X\}$. It is easy to prove that J_t, J_D, J_f, J_c are topological spaces on X, J_t is known as the trivial or indiscrete topology on X, J_D is known as the discrete topology on X, J_f is known as the cofinite topology on X, J_c is known as the co-countable topology on X.

Recall that a set A is a countable set if and only if A is a finite set or A is a countably infinite set. Also note that A is a countably infinite set if and only if there exists a bijective function f from $\mathbb{N}$ to A, where $\mathbb{N}$ is the set of all natural numbers. Also it is known that a nonempty set A is a countable set if and only if there exists a surjective function say $f : \mathbb{N} \to A$.

Now let us prove that $J_c = \{A : X \setminus A$ is a countable subset of A or $A^c = X\}$ is a topology on X.

Proof: Now $\phi^c = X$ implies $\phi \in J_c$, $X^c = \phi$ and ϕ is a countable set implies $X \in J_c$.

Hence

$$\phi, X \in J_c. \qquad (1)$$

Let J be a nonempty set and for each $\alpha \in J$, $A_\alpha \in J_\alpha$.

Claim: $\bigcup_{\alpha \in J} A_\alpha \in J_c$.

Now $\left(\bigcup_{\alpha\in J} A_\alpha\right)^c = \bigcap_{\alpha\in J} A_\alpha^c$. Hence we will have to prove that either $\bigcap_{\alpha\in J} A_\alpha^c = X$ or $\bigcap_{\alpha\in J} A_\alpha^c$ is a countable subset of X. If $\bigcap_{\alpha\in J} A_\alpha^c = X$ then we are through (from (T1)). Suppose not. This implies for at least one $\alpha_0 \in J$, $A_{\alpha_0}^c \neq X$, $A_{\alpha_0} \in J_c$ implies $A_{\alpha_0}^c$ is a countable set. Since $\bigcap_{\alpha\in J} A_\alpha^c \subseteq A_{\alpha_0}^c$, $\bigcap_{\alpha\in J} A_\alpha^c$ is a countable set (subset of a countable set is countable). We have proved that either $\bigcap_{\alpha\in J} A_\alpha^c = X$ or $\bigcap_{\alpha\in J} A_\alpha^c$ is a countable set. Hence

$$\bigcup_{\alpha\in J} A_\alpha \in J_c. \tag{2}$$

Let $A_1, A_2 \in J_c$ implies that A_1^c is a countable set or $A_1^c = X$ and $A_2 \in J_c$ implies A_2^c is a countable set or $A_2^c = X$. Now $(A_1 \cap A_2)^c = A_1^c \cup A_2^c = X$ when A_1^c or $A_2^c = X$ or $A_1^c \cup A_2^c$ is a countable set since in this case both A_1^c and A_2^c are countable sets. Hence $(A_1 \cap A_2)^c = X$ or it is a countable set. This implies that

$$A_1 \cap A_2 \in J_c. \tag{3}$$

From Eqs. (1), (2), and (3), J_c is a topology on X. Now let us give some examples to illustrate the natural way of obtaining topologies once we know bases satisfying (B1) and (B2).

Example: Let X be a nonempty set and $\mathscr{B} = \{\{x\} : x \in X\}$. Then $\mathscr{B}$ is a basis for a topology on X.

(i) For every $x \in X$ there exists $B = \{x\} \in \mathscr{B}$ such that $x \in B$.

(ii) $B_1, B_2 \in \mathscr{B}$ and $x \in B_1 \cap B_2$ implies there exists $B_3 = \{x\} \in \mathscr{B}$ such that $x \in B_3 \subseteq B_1 \cap B_2$. Hence both (B1) and (B2) are satisfied. This implies that the collection $\mathscr{B} = \{\{x\} : x \in X\}$ is a basis for a topology on X.

Now let us find out $J_{\mathscr{B}}$, the topology, generated by $\mathscr{B}$. In theorem 1 we have proved that if we define $J_{\mathscr{B}}$ as $J_{\mathscr{B}} = \{U \subseteq X : x \in U$ implies there exists $B \in \mathscr{B}$ such that $x \in B \subseteq U\}$ is a topology on X.

In this case for any nonempty subset U of X, $x \in U$ implies there exists $B = \{x\}$ such that $x \in B \subseteq U$. Hence by the definition of $J_{\mathscr{B}}$, $A \in \mathscr{B}$ whenever A is a nonempty subset of X. Also the null set $\phi \in J_{\mathscr{B}}$ (recall the proof given in theorem 1). Hence A ⊆ X implies $A \in J_{\mathscr{B}}$ implies $\mathcal{P}(X) \subseteq J_{\mathscr{B}}$. Also by the definition, $J_{\mathscr{B}} \subseteq \mathcal{P}(X)$, the collection of all subsets of X. This implies that $J_{\mathscr{B}}$ is same as the discrete topology J_D defined on X.

Exercise: Let X be a nonempty set and let d be a metric on X. That is (X, d) is a given metric space. Then prove that the collection $\mathscr{B}$ defined as $\mathscr{B} = \{B(x, r) : x \in X, r > 0\}$ is a basis for J_D.

Now let us consider the special case $X = \mathbb{R}$, the set of all real numbers and $d(x, y) = |x - y|$ for $x, y \in \mathbb{R}$. Then d is a metric on $\mathbb{R}$. What is the collection $\mathscr{B} = \{B(x,r) : x \in X, r > 0\}$. Note that $B(x, r) = (x - r, x + r) = (a,b)$, where $a = x - r \in \mathbb{R}$, $b = x + r \in \mathbb{R}$ with a < b. That is $\mathscr{B} \subseteq \{(a, b) : a, b \in \mathbb{R}, a < b\} = \mathscr{B}'$.

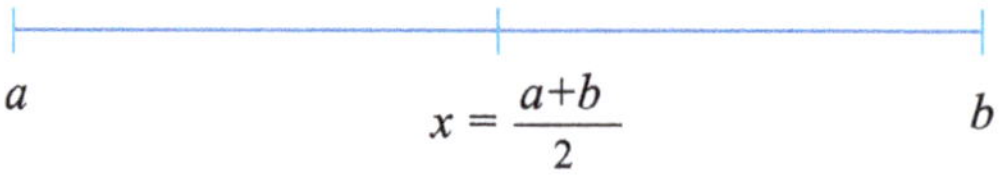

On the other hand take a member say $B \in \mathscr{B}'$. Since $B \in \mathscr{B}'$ there exist $a, b \in \mathbb{R}$, $a < b$ such that B = (a, b). Now let $x = \dfrac{a+b}{2}$ and $r = \left|\dfrac{a-b}{2}\right| = \dfrac{b-a}{2} > 0$. Then $B(x, r) = (x - r, x + r) = \left(\dfrac{a+b}{2} - \dfrac{b-a}{2}, \dfrac{a+b}{2} + \dfrac{b-a}{2}\right) = (a,b)$ implies $(a, b) = B(x, r) \in \mathscr{B}$ implies $\mathscr{B}' \subseteq \mathscr{B}$. Also we have $\mathscr{B} \subseteq \mathscr{B}'$ and hence $\mathscr{B} = \{B(x, r) : x \in \mathbb{R}, r > 0\} = \{(a, b) : a, b \in \mathbb{R}, a < b\} = \mathscr{B}'$. That is $\{(a,b) : a,b \in \mathbb{R}, a < b\}$ is basis for a topology on $\mathbb{R}$ and $J_{\mathscr{B}} = J_d$. This topology is called the standard or usual topology on $\mathbb{R}$, and it is denoted by J_s.

Exercises: (i) Prove that $\mathscr{B}_{\mathbb{Q}} = \{(a,b) : a,b \in \mathbb{Q}, a < b\}$, where $\mathbb{Q}$ - the set of all rational numbers is also a basis for the usual topology on $\mathbb{R}$. That is $J_{\mathscr{B}_{\mathbb{Q}}}$ is same as the usual topology on $\mathbb{R}$.

(ii) Is $\mathscr{B}_0 = \left\{B\left(x, \dfrac{1}{n}\right) : x \in \mathbb{Q}, \text{n} \in \mathbb{N}\right\}$ a basis for the usual topology on $\mathbb{R}$? Justify your answer.

(iii) It is given that (X, d) is a metric space. Now prove that $\mathscr{B}' = \left\{B\left(x, \dfrac{1}{n}\right) : x \in N, \text{n} \in \mathbb{N}\right\}$ is a basis for a topology on X. Also prove that $J_{\mathscr{B}'} = J_d$.

Definition: A subset A of a topological space (X, d) is said to be a closed set if the complement X\A = A^c of A is an open set.

Use the DeMorgan's law to prove the following theorem.

Theorem 3. In a topological space (X, J) we have:

(i) X and ϕ are closed.

(ii) Suppose we have a nonempty index set J and to each α ∈ J, A_α is a closed subset of X. Then $\bigcap_{\alpha \in J} A_\alpha$ is a closed subset of X. That is arbitrary intersection of closed sets is closed.

(iii) If A_1, A_2 are closed sets then $A_1 \cup A_2$ is also a closed set.

Use induction to prove that finite union of closed sets is closed.

Now let us prove the following theorem which tells us when a subcollection $\mathscr{B}$ of a given topology J on X generates the topology J.

Theorem 4. Let (X, J) be a topological space and $\mathscr{B} \subseteq J$. Further suppose for each $A \subseteq J$ and $x \in A$ there exists $B \in \mathscr{B}$ such that $x \in B \subseteq A$. Then $\mathscr{B}$ is a basis for a topology on X and $J_{\mathscr{B}} = J$.

Proof: First let us prove that $\mathscr{B}$ is a basis for a topology on X.

(B1) Let $x \in X$. Since $X \in J$, by hypothesis, there exists $B \in \mathscr{B}$ such that $x \in B$.

(B2) Let $B_1, B_2 \in \mathscr{B}$ and $x \in B_1 \cap B_2$. It is given that $\mathscr{B} \subseteq J$. Hence $B_1, B_2 \in J$ and this implies $B_1 \cap B_2 \in J$. Now $x \in B_1 \cap B_2$ and $B_1 \cap B_2 \in J$ implies there exists $B_3 \in \mathscr{B}$ such that $x \in B_3 \subseteq B_1 \cap B_2$. From (B1) and (B2) we see that $\mathscr{B}$ is a basis for a topology on X.

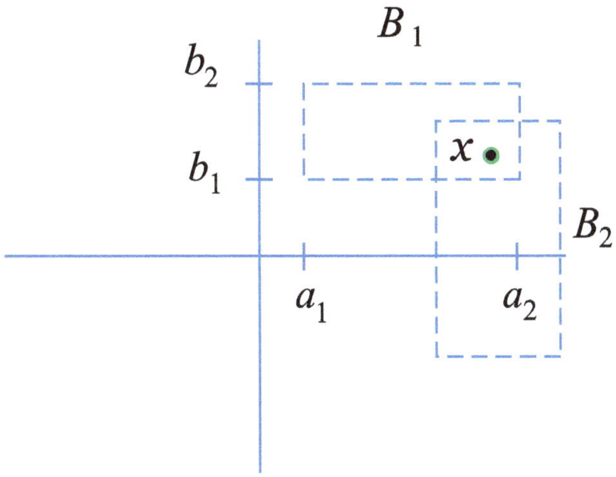

Now let us prove that $J_{\mathscr{B}} = J$. Let $U \in J_{\mathscr{B}}$.

Claim: $U \in J$. If $U = \phi$ then this implies that $U \in J$. Otherwise let $x \in U$. By the definition of $J_{\mathscr{B}}$ there exists $B_x \in \mathscr{B} \subseteq J$ such that $x \in B_x \subseteq U$. This implies that $\bigcup_{x \in U} B_x = U$ and hence $U \in J$.

Conversely, let $\phi \neq U \in J$. Then for each $x \in U$ there exists $B \in \mathscr{B} \subseteq J_{\mathscr{B}}$ such that $x \in B \subseteq U$. This proves that $\bigcup_{x \in U} B_x = U \in J_{\mathscr{B}}$. Hence $J \subseteq J_{\mathscr{B}}$. Already we have proved that $J_{\mathscr{B}} \subseteq J$ and therefore $J = J_{\mathscr{B}}$.

Now it is natural to introduce the following definition.

Definition: If $(X, \mathcal{J})$ is a topological space and $\mathscr{B} \subseteq \mathcal{J}$ such that for each $A \in \mathcal{J}$ and $x \in A$ there exists $B \in \mathscr{B}$ such that $x \in B \subseteq A$, then we say that $\mathscr{B}$ is a basis for $\mathcal{J}$.

Theorem 5. Let $(X, \mathcal{J})$ be a topological space and $\mathscr{B} \subseteq \mathcal{J}$. Then $\mathscr{B}$ is a basis for $\mathcal{J}$ if and only if every member A of $\mathcal{J}$ is the union of member of some subcollection of $\mathscr{B}$.

To have a feeling of this concept do the following exercise:

Exercise: For $a_1, a_2, b_1, b_2 \in \mathbb{R}$, $a_1 < a_2$, $b_1 < b_2$ *let* $R = \{(x_1, x_2) \in \mathbb{R}^2 : a_1 < x_1 < a_2,\ b_1 < x_2 < b_2\}$. That is, R is an open rectangle having sides parallel to the coordinate axes. Let $\mathscr{B}$ be the collection of all such open rectangles. Now it is easy to see that $\mathscr{B}$ is a basis for the topology $\mathcal{J}_d$ on $\mathbb{R}^2$, where d is the Euclidean metric on $\mathbb{R}^2$.

Remark: Let X be a set and $\mathcal{S}$ be a collection of subsets of X. Suppose $\bigcup_{A \in \mathcal{S}} A = X$. Then we say that $\mathcal{S}$ is a subbasis for a topology on X. In this case, let $\mathscr{B} = \{A \subseteq X : A = \bigcap_{B \in \mathcal{F}} B$, for a finite subcollection $\mathcal{F}$ of $\mathcal{S}\}$. Then it is easy to prove that $\mathscr{B}$ is a basis for a topology on X. The topology $\mathcal{J}_{\mathscr{B}}$ generated by $\mathscr{B}$ is called the topology on X generated by the subbasis $\mathcal{S}$.

Exercises: (i) Let $\mathcal{S}_1 = \{(a, \infty) : a \in \mathbb{R}\}$. Prove that $\mathcal{S}_1$ is a subbasis for a topology on $\mathbb{R}$. Find out the topology $\mathcal{J}_1$ generated by $\mathcal{S}_1$.

(ii) Let $\mathcal{S}_2 = \{(-\infty, a) : a \in \mathbb{R}\}$. Prove that $\mathcal{S}_2$ is a subbasis for a topology on $\mathbb{R}$. Find out the topology $\mathcal{J}_2$ generated by $\mathcal{S}_2$.

Interior Points, Limit Points, Boundary Points, Closure of a Set

Let A be a nonempty subset of a topological space $(X, \mathcal{J})$ and $x \in A$. Then x is said to be an interior point of A if there exists an open set U such that $x \in U$ and U ⊆ A. Also the collection of all interior points of A denoted it by int(A) or A°. For a nonempty subset A of a topological space $(X, \mathcal{J})$, a point $x \in X$ is said to be a limit point or an accumulation point of A if for each open set U containing x, $U \cap (A \setminus \{x\}) \neq \phi$.

For A ⊆ X, the derived set of A denoted by A' is defined as $A' = \{x \in X : x$ is a limit point of A}.

A point $x \in X$ is said to be a boundary point of A if for each open set U containing $x, U \cap A \neq \phi$ and $U \cap A^c \neq \phi$.

For A ⊆ X, the boundary of A, denoted by bd(A), is defined as $bd(A) = \{x \in X :$ for each open set U containing x, $U \cap A \neq \phi$ and $U \cap A^c \neq \phi\}$. That is bd(A) is the collection of all boundary points of A.

For A ⊆ X, the closure of A denoted by $\overline{A}$ or cl(A), is defined as $\overline{A} = A \cup A'$.

Examples: (i) Let X = {1, 2, 3} and $J = \{\phi, X, \{1\}, \{2\}, \{1, 2\}, \{1, 3\}, \{2, 3\}\}$. Is J a topology on X? Let A = {1, 3}, B = {2, 3}. Here $A \in J$, $B \in J$, but $A \cap B = \{3\} \notin J$. Hence J is not a topology on X.

(ii) Let X = {1,2,3} and $J = \{\phi, X, \{1\}, \{2\}, \{1, 2\}\}$ then J is a topology on X. Now A = {2, 3} is a subset of X. 2 ∈ A and also there is an open set U = {2} such that 2 ∈ U and U ⊆ A. Hence 2 is an interior point of A. But 3 is not an interior point of A. How to check 3 is an interior point of A or not?

Step 1: First check whether 3 ∈ A (if x is an interior point of A then it is essential that $x \in A$). Yes here 3 ∈ {2, 3} = A.

Step 2: Now find out all the open sets containing 3. X is the only open set containing 3. But this open set is not contained in A. Hence 3 is not an interior point of A. What will happen if the given set A is an open subset of a topological space X. Our aim is to check whether an element $x \in X$ is an interior point of A.

Step 1: It is essential that $x \in A$.

Step 2: Is it necessary to find out all the open sets containing x? Of course not necessary. It is enough if we find at least an open set U such that $x \in U$ and U ⊆ A. In this case the given set A is an open set and hence there exists an open set U = A such that $x \in U$ and U = A ⊆ A. Therefore every element x of A is an interior point of A. That is $A \subseteq A^{\circ}$. By definition $A^{\circ} \subseteq A$. Hence $A^{\circ} = A$. That is if A is an open set then $A^{\circ} = A$. What about the converse? Suppose for a subset A of X, $A^{\circ} = A$. Is A an open set? Yes, A is an open subset of X. Take $x \in A$. Then $x \in A^{\circ}$. Hence by the definition of A° there exists at least one open set say U_x such that $x \in U_x$ and $U_x \subseteq A$. This implies that $A = \bigcup_{x \in A} U_x$. Now by the definition, J is closed under arbitrary union. Hence for each $x \in A$, $U_x \in J$ implies $\bigcup_{x \in A} U_x \in J$ implies $A \in J$. That is, A is an open set. Thus, we have proved:

Theorem 6. For a subset A of topological space (X, J), A is open if and only if $A^{\circ} = A$.

Now let us prove that for any subset A of X, A° is an open set and if B is an open set contained in A (B ⊆ A) then $B \subseteq A^{\circ}$.

Theorem 7. For any subset A of a topological space (X, J), A° is the largest open set contained in A.

Proof: If $A = \phi$ then $A^{\circ} = \phi$. For $A \neq \phi$, let us prove that $B = A^{\circ}$ is an open set. Due to theorem 6, it is enough to prove that $\left(A^\circ\right)^\circ = A^{\circ\circ} = A^\circ$. *If* $A^\circ = \phi$ then $A^{\circ\circ} = \phi$ and we are through. Also by definition $A^{\circ\circ} \subseteq A^\circ$. Let $x \in A^\circ$. Then by the definition, there exists an open set U_x such that $x \in U_x \subseteq A$. Note that for each $y \in U_x$, $y \in U_x \subseteq A$. That is y ∈ A and there exists an open set U_x such that $y \in U_x$ and $U_x \subseteq A$. This implies that $y \in A^\circ$. That is $y \in U_x$ implies $y \in A^\circ$ implies $U_x \subseteq A^\circ$. We have the following:

- $x \in A^\circ$ and,
- there exists an open set U_x such that $x \in U_x$, $U_x \subseteq A^\circ$.

This implies that $x \in A^{\circ\circ}$. That is $x \in A^\circ$ implies $x \in A^{\circ\circ}$ implies $A^\circ \subseteq A^{\circ\circ}$. Also we have $A^{\circ\circ} \subseteq A^\circ$ implies $(A^\circ)^\circ = A^\circ$. From the theorem 6, A° is an open set. Also by definition, $A^\circ \subseteq A$.

To prove the second part assume that B is an open subset of X such that B ⊆ A. Now we aim to prove $B \subseteq A^\circ$. Which is obvious since for each $x \in B$ there exists an open set B such that $x \in B$ and B ⊆ A. Hence by definition $B \subseteq A^\circ$.

Consider $X = \{1, 2, 3\}$, $J = \{\phi, X, \{1\}, \{1, 2\}\}$ and A = {1, 2}. What is A', the collection of all limit points of A. Is $1 \in A'$? The answer is no. Since {1} is an open set containing 1, but $\{1\} \cap A \setminus \{1\} = \{1\} \cap \{2\} = \phi$. Is $2 \in A'$? Again the answer is no. Since {2} is an open set containing 2 and $\{2\} \cap A \setminus \{2\} = \{2\} \cap \{1\} = \phi$. Is $3 \in A'$? First find out all the open sets containing 3.

Here the whole space X is the only open set containing 3 and $X \cap A \setminus \{3\} =$ $= \{1,2,3\} \cap \{1,2\} \neq \phi$. That is for each open set U containing 3, the condition namely $U \cap A \setminus \{3\} \neq \phi$ is satisfied. Hence 3 is a limit point of A. That is 3 ∈ A'. Here $A' = \{3\}$. What is $\bar{A}$, the closure of A? By definition $\bar{A} = A \cup A' = \{1, 2\} \cup \{3\} = \{1, 2, 3\}$.

Now let us prove that for any subset A of a topological space X,

- $\bar{A}$ is a closed set and $A \subseteq \bar{A}$.
- Whenever B is a closed set such that A ⊆ B then $\bar{A} \subseteq B$ that is we aim to prove:

Theorem 8. For a subset A of a topological space X, $\bar{A}$ is always a closed set containing A and it is the smallest closed set containing A.

Proof: Let us prove that $\left(\bar{A}\right)^c = X \setminus \bar{A}$ is an open set. Hence we will have to prove that interior of $\left(\bar{A}\right)^c$ is itself. Let $x \in \left(\bar{A}\right)^c$ then $x \notin \bar{A}$. Hence there exists an open set U containing x such that $U \cap A = \phi \Rightarrow U \subseteq A^c$. This imply that x is an interior point of A^c, but we have to prove that x is an interior point of $\left(\bar{A}\right)^c$. So it is enough to prove that $U \subseteq \left(\bar{A}\right)^c$.

Suppose not. Then there exists y ∈ U such that $y \notin \left(\bar{A}\right)^c$ implies $y \in \left(\bar{A}\right)$. Also U is an open set containing y. Hence $U \cap A \neq \phi$. This is contradiction to $U \cap A = \phi$. We arrived at this contradiction by assuming that U is not a subset of$\left(\overline{A}\right)^c$. Hence $U \subseteq \left(\overline{A}\right)^c$, where U is an open set containing x and $x \in \left(\overline{A}\right)^c$. Therefore every point of $\left(\overline{A}\right)^c$ is an interior point. This implies that $\left(\overline{A}\right)^c$ is an open set and hence $\overline{A}$ is a closed set.

Now let B be a closed set containing A then we will have to prove that $\overline{A} \subseteq B$. That is to prove $\overline{A} \cap B^c = \phi$.

Suppose not. Then there exists $x \in \overline{A} \cap B^c$, B^c is an open set containing x. Now if $x \in \overline{A} = A \cup A'$ is such that $x \in A$ then $x \in B$ (given that A ⊆ B) and we are through. On the other hand if $x \in A'$ and $x \notin A$ then by the definition of A', $B^c \cap A \setminus \{x\} = B^c \cap A \neq \phi$ (note: $x \notin A \Rightarrow A \setminus \{x\} = A$) is a contradiction since A ⊆ B implies $A \cap B^c \subseteq B \cap B^c = \phi$. Hence $\overline{A} \cap B^c = \phi$. That is $\overline{A} \subseteq B$.

Hausdorff Topological Spaces

Definition: A topological space (X, J) is said to be a Hausdorff topological space (or Hausdorff space) if for $x, y \in X$, $x \neq y$, there exist U, $V \in J$ such that $(i) x \in U, y \in V$, $(ii) U \cap V = \phi$.

Note. In definition, in place of if it is also absolutely correct to use if and only if. That is, definition can also be read as:

A topological space (X, J) is said to be a Hausdorff topological space (or Hausdorff space) if and only if (iff) for $x, y \in X$, $x \neq y$, there exist U, $V \in J$ such that $(i) x \in U, y \in V$, $(ii)\ U \cap V = \phi$.

What is important to note here (that is while giving a definition) is one can use interchangeably "if" and "if and only if".

Example: If $X = \mathbb{R}$, and J_s is the standard topology on $\mathbb{R}$, then $(\mathbb{R}, J_s)$ is a Hausdorff space.

Example: Every discrete topological space (X, J) is a Hausdorff space.

Example: If X is a set containing at least two elements and $J = \{\phi, X\}$ then (X, J) is not a Hausdorff space.

Example: If $X = \mathbb{R}$, $\mathscr{B} = \{(a, \infty) : a \in \mathbb{R}\}$ then $\mathscr{B}$ is a basis for a topology $J_{\mathscr{B}}$ on $\mathbb{R}$. It is easy to see that $(\mathbb{R}, J_{\mathscr{B}})$ is not a Hausdorff space.

Example: $\mathscr{B}_l = \{[a, b) : a, b \in \mathbb{R},\ a < b\}$, $J_l = J_{\mathscr{B}_l}$ is known as the lower limit topology on $\mathbb{R}$ and $J_s \subseteq J_l$. Hence $(\mathbb{R}, J_l)$ is a Hausdorff space.

Note. Weaker topology is Hausdorff implies stronger is also Hausdorff.

Let X be an infinite set and J_f be the cofinite topology on X. Also let $x, y \in X, x \neq y$. If $U \in J_f$ and $x \in U$ then U^c is finite, because $U^c \neq X$. Also $y \in V \in J_f$ implies V^c is finite. If $U \cap V = \phi$, then $X = (U \cap V)^c = U^c \cup V^c$ and hence X is a finite set. Which gives a contradiction. Therefore $U \cap V \neq \phi$. Hence J_f is not a Hausdorff space.

Example: Let X = {a, b, c} and $J = \{\phi, X, \{a\}, \{b\}, \{a, b\}\}$. Then (X, J) is not a Hausdorff space.

Example: $\mathscr{B} = \{U_1 \times U_2 \times \cdots \times U_n \times \mathbb{R} \times \mathbb{R} \times \cdots$: each U_i is open in $\mathbb{R}$, $i = 1, 2, \ldots, n, n \in \mathbb{N}\}$ is a basis for a topology J (known as product topology) on $\mathbb{R}^w$ where $\mathbb{R}^w = \left\{x = (x_n)_{n=1}^{\infty} : x_n \in \mathbb{R} \forall n\right\}$. Now $X = \mathbb{R}^w$, $x = (x_n) \in X$ and $y = (y_n) \in X$ such that $x \neq y$. Therefore there exists $k \in \mathbb{N}$ such that $x_k \neq y_k$. Let $\epsilon = \frac{|x_k - y_k|}{2} > 0$ and $U_k = (x_k - \epsilon, x_k + \epsilon)$, $V_k = (y_k - \epsilon, y_k + \epsilon)$.

Let $U = \mathbb{R} \times \mathbb{R} \times \cdots \times \mathbb{R} \times U_k \times \mathbb{R} \times \mathbb{R} \cdots$ and $V = \mathbb{R} \times \mathbb{R} \times \cdots \times \mathbb{R} \times V_k \times \mathbb{R} \times \mathbb{R} \cdots$.

Clearly, $x \in U$, $y \in V$ and $U \cap V = \mathbb{R} \times \mathbb{R} \times \cdots \times \mathbb{R} \times \phi \times \mathbb{R} \times \mathbb{R} \times \cdots = \phi$. Hence $X = \mathbb{R}^w$ is a Hausdorff space.

Note. $\prod_{n=1}^{\infty} \left(\frac{-1}{n}, \frac{1}{n} \right)$ is not an open set in the product topology on $\mathbb{R}^w$.

Definition: A sequence $\{x_n\}$ in a topological space (X, J) is said to converge to a point $x \in X$ if for each open set U containing x there exists $n_0 \in \mathbb{N}$ such that $x_n \in U, \forall n \geq n_0$. In symbol we write $x_n \to x$ as $n \to \infty$.

Note that $x_n \to x$ as $n \to \infty$ if and only if for each open set U containing x there exists $n_0 \in \mathbb{N}$ such that $x_n \in U, \forall n \geq n_0$.

Example: If $X \neq \phi$, $J = \{\phi, X\}$, and $\{x_n\}$ is a sequence in X. Then $\{x_n\}$ converges to every element of X.

Example: If X be an infinite set, $J_f = \{A \subseteq X : A^c \textit{ is finite or } A^c = X\}$, then J_f is not Hausdorff. Let $\{x_n\}$ be a sequence in X and $x \in X$. Now $U \in J_f$ and $x \in U \Rightarrow U^c$ is finite. If $U^c = \phi$ then U = X. Otherwise U^c is nonempty and finite and hence $J = \{n : x_n \in U^c\}$ is a finite set. If $J = \phi$ let $n_0 = 1$, otherwise let $n_0 = \max\{n : n \in J\}$. Then $x_n \in U, \forall n > n_0$. Therefore $x_n \to x$ as $n \to \infty$. So, (x_n) converges to every element of X.

Theorem 9. Let (X, J) be a Hausdorff space and let $A \subseteq X$. Then an element $x \in A'$ if and only if for each open set U containing x, $U \cap A$ is an infinite set.

Proof: Assume that $x \in A'$ and suppose for some open set U containing x, $U \cap (A \setminus \{x\})$ is a nonempty finite set. Let $U \cap (A \setminus \{x\}) = \{x_1, x_2, \ldots, x_n\}$. For each i, $x_i \neq x$ and (X, J) is a Hausdorff space implies there exist open sets U_i and V_i such that

$x_i \in U_i$, $x \in V_i$ and $U_i \cap V_i = \phi$. Note that $x_i \notin V_i$ for all $i = 1, 2, \ldots, n$ and $x \in V = \bigcap_{i=1}^{n} V_i$, V is an open set. Also $x \in U$. Therefore $x \in U \cap V$. But $(A \setminus \{x\}) \cap (U \cap V) = \phi$ which is a contradiction. Hence U ∩ A is an infinite set.

Conversely, if for $x \in X$, U ∩ A is an infinite set for each open set containing x then in particular $U \cap (A \setminus \{x\}) \neq \phi$, for each open set containing x. Hence x is a limit point of A. That is $x \in A'$.

Exercise: Let (X, J) be a topological space such that for each x in X, $\{x\}$ is closed in X. Then prove that an element $x \in A'$ if and only if for each open set U containing x, U ∩ A is an infinite set.

Note. If X is a Hausdorff space, and if A is a finite subset of X, then $A' = \phi$.

Definition: A topological space (X, J) is said to be metrizable if there exists a metric d on X such that $J_d = J$.

Theorem 10. Let (X, d) be a Hausdorff topological space. Then $\{x\}$ is closed for each $x \in X$.

Proof: If $\{x\}^c = \phi$, then $X = \{x\}$ is a closed set. Suppose $\{x\}^c \neq \phi$, this implies that for each $y \in \{x\}^c$, $y \neq x$. Now (X, J) is a Hausdorff space implies there exist open sets U_x, V_y in X such that $x \in U_x$, $y \in U_y$ and $U_x \cap V_y = \phi$. Therefore $V_y \subseteq \{x\}^c$ implies $y \in \left(\{x\}^c\right)^o$. This implies that $\{x\}$ is closed.

Theorem 11. Let (X, J) be a topological space, Y ⊆ X and let $J_Y = \{A \cap Y : A \in J\}$ then J_Y is a topology on Y.

Proof: (i) $\phi \in J_Y$. Now $\phi \in J_X$ implies $\phi \cap Y = \phi \in J_Y$.

(ii) Since $X \in J_X \Rightarrow X \cap Y = Y \in J_Y$.

Let $A_i \in J_Y$ for $i \in I$. Now $A_i \in J_Y$ implies there exists $B_i \in J = J_X$ such that $A_i = B_i \cap Y$. Now $B_i \in J$ for each $i \in I$, J is a topology on X implies $\bigcup_{i \in I} B_i \in J$. Hence $\bigcup_{i \in I} A_i = \left(\bigcup_{i \in I} B_i\right) \cap Y$ is open in J_Y.

(iii) Let $A_1, A_2, \ldots, A_n \in J_Y$. Then there exists $B_i \in J_X$ such that $A_i = B_i \cap Y$. Therefore $\bigcap_{i=1}^{n} B_i \in J_X$. Therefore $A_1 \cap A_2 \cap \cdots \cap A_n = (B_1 \cap Y) \cap (B_2 \cap Y) \cap \cdots \cap (B_n \cap Y) = \left(\bigcap_{i=1}^{n} B_i\right) \cap Y \in J_Y$, and this implies that $A_1 \cap A_2 \cap \cdots \cap A_n \in J_Y$. From (i), (ii), and (iii) J_Y is a topology on Y.

Definition: Let (X, J) be a topological space, and let $J_Y = \{A \cap Y : A \in J\}$ then

J_Y is a topology on Y . This topology J_Y is called the relative topology on Y induced by J .

Note. Let $A \subseteq Y \subseteq X$. We use $\overline{A}$ to denote the closure of A in X, and $\overline{A}_Y$ to denote the closure of A in Y.

Result: For $A \subseteq Y$, $\overline{A}_Y = \overline{A} \cap Y$.

Proof: It is always true that $\overline{A}_Y \subseteq \overline{A}$. Now $\overline{A}_Y$ = closure of A in Y , $\overline{A}_Y \subseteq Y$ and hence $\overline{A}_Y \subseteq \overline{A} \cap Y$. Let $x \in \overline{A} \cap Y$. This implies that $x \in \overline{A}$ and $x \in Y$. Then for each open set U containing x, $U \cap A \neq \phi$. Hence $(U \cap Y) \cap A = U \cap A \neq \phi$. Thus $x \in \overline{A}_Y$.

Therefore $\overline{A} \cap Y = \overline{A}_Y$.

Theorem 12. Let (X, J) be a topological space and A, B be subsets of X. Then

(i) $\overline{A \cup B} = \overline{A} \cup \overline{B}$, (ii) $\overline{A \cap B} \subseteq \overline{A} \cap \overline{B}$, (iii) $\overline{\overline{A}} = \overline{A}$, (iv) $(A \cup B)^\circ \supseteq A^\circ \cup B^\circ$,

(v) $(A \cap B)^\circ = A^\circ \cap B^\circ$, (vi) $(A^\circ)^\circ = A^\circ$.

Proof: (i) An element $x \in \overline{A}$ if and only if for each open set U containing x such that $U \cap A \neq \phi$. Let $x \in \overline{A \cup B}$. This implies that for every neighbourhood U containing x,

$$U \cap (A \cup B) \neq \phi \Rightarrow (U \cap A) \cup (U \cap B) \neq \phi. \tag{4}$$

Suppose $x \notin \overline{A}$ and $x \notin \overline{B}$. Then for some open sets U, V containing x such that $U \cap A = \phi$ and $V \cap B = \phi$. Let $U_0 = U \cap V$. Then $U_0 \cap (A \cup B) = (U_0 \cap A) \cup (U_0 \cap B) = \phi$, a contradiction to Eq. (4). Therefore $x \in \overline{A}$ or $x \in \overline{B}$. This proves that

$$\overline{A \cup B} \subseteq \overline{A} \cup \overline{B}. \tag{5}$$

Now let $x \in \overline{A} \cup \overline{B}$. This implies that $x \in \overline{A}$ or $\overline{B}$ or both. Hence for each neighbourhood U of x, $U \cap A \neq \phi$ or $U \cap B \neq \phi$. This implies that $(U \cap A) \cup (U \cap B) \neq \phi$ implies $U \cap (A \cup B) \neq \phi$. This shows that $x \in \overline{A \cup B}$. Therefore

$$\overline{A} \cup \overline{B} \subseteq \overline{A \cup B}. \tag{6}$$

From Eqs.(5) and (6) $\overline{A \cup B} = \overline{A} \cup \overline{B}$.

(ii) Let $x \in \overline{A \cap B}$. Then for every open set U containing x, $U \cap (A \cap B) \neq \phi$. Hence $U \cap A \neq \phi$ and $U \cap B \neq \phi$. This implies $x \in \overline{A}$ and $x \in \overline{B}$. Hence $\overline{A \cap B} \subseteq \overline{A} \cap \overline{B}$.

(iii) $\overline{A}$ is the smallest closed set containing A implies $\overline{A} \subseteq \overline{\overline{A}}$. Also $\overline{A}$ is a closed set containing $\overline{A}$. But $\overline{\overline{A}}$ is the smallest closed set containing $\overline{A}$. Hence $\overline{\overline{A}} \subseteq \overline{A}$. So we have $\overline{A} = \overline{\overline{A}}$.

(iv) Now $x \in A^\circ \cup B^\circ \Rightarrow x \in A^\circ$ or $x \in B^\circ$ or both. Without loss of generality assume that $x \in A^\circ$. Then there exists $U \in J$ such that $x \in U \subseteq A$. This implies that $x \in U \subseteq (A \cup B)$. Hence $x \in (A \cup B)^\circ$. That is $x \in A^\circ \cup B^\circ$ implies $x \in (A \cup B)^\circ$. Hence $(A \cup B)^\circ \subseteq A^\circ \cup B^\circ$.

(v) Let $x \in (A \cap B)^\circ$. Then there exists $U \in J$ such that $x \in U \subseteq A \cap B$. This implies $x \in A^\circ$ and $x \in B^\circ$. Hence $A^\circ \cap B^\circ \subseteq (A \cap B)^\circ$.

Now let $x \in A^\circ \cap B^\circ$. Then $x \in A^\circ$ and $x \in B^\circ$. Hence there exists $U \in J$ such that $x \in U \subseteq A$ and $V \in J$ such that $x \in V \subseteq B$, hence $x \in U \cap V \subseteq A \cap B$. Hence $x \in (A \cap B)^\circ$. This implies $(A \cap B)^\circ = A^\circ \cap B^\circ$.

(vi) Now $A^{\circ\circ}$ is the largest open set contained in A° implies $A^{\circ\circ} \subseteq A^\circ$. Also A° is an open set contained in A°. Hence $A^\circ \subseteq A^{\circ\circ}$. So we have proved that $A^\circ = A^{\circ\circ}$.

Example: For $A = \mathbb{Q}$, $B = \mathbb{Q}^c$, $A \cup B = \mathbb{R}$ and $(A \cup B)^\circ = \mathbb{R}$, $A^\circ = \phi, B^\circ = \phi$, $\mathbb{R} \neq \phi$. Hence $(A \cup B)^\circ \neq A^\circ \cup B^\circ$.

$\overline{A} = \mathbb{R}$, $\overline{B} = \mathbb{R}$, $\overline{A \cap B} = \overline{\phi} = \phi$. $\overline{A} \cap \overline{B} = \mathbb{R}$. Hence $\overline{A} \cap \overline{B} \neq \overline{A \cap B}$.

Definition: Let (X, J) be a topological space. Then a set $B \subseteq Y$ is open in Y if and only if $B = A \cap Y$, for some $A \in J$.

Example: The set [0, 1) is open in Y = [0,∞). Note that A = (−1, 1)∩[0,∞) = [0, 1). Now $A^c = Y \setminus A = [1, \infty)$ is open in (Y, J_Y) if and only if for each $x \in [1, \infty)$ there exists $U \in J_Y$ such that $x \in U \subseteq [1, \infty)$. But 1 is not an interior point of A^c. Hence A^c is not open and therefore A is not closed. Whereas [1,∞) is open in $[1, \infty) \cup \mathbb{Z}$.

Exercises: (i) Let $X = \mathbb{R}$ and J_f be the cofinite topology on $\mathbb{R}$. Let $Y = \mathbb{Q}$ what is $J_f / Y = ?$

(ii) Prove that for each $x \in \mathbb{R}$, the sequence $(x_n) = \left(\frac{1}{n}\right) \to x$ *in* $(\mathbb{R}, J_f)$.

Result: Let (X, J) be a topological space and $\mathscr{B}$ is a basis for J then for each $Y \subseteq X$, $\mathscr{B}_Y = \{B \cap Y : B \in \mathscr{B}\}$ is a basis for J_Y.

Proof: Let $U \in J_Y$ and $x \in U$. $U \in J_Y$ implies $U = V \cap Y$ for some $V \in J$. Now $x \in V$, $V \in J$ and $\mathscr{B}$ is a basis for J this implies that there exists $B \in \mathscr{B}$ such that $x \in B \subseteq U$. Therefore $B \cap Y \subseteq V \cap Y = U$. Now $B \cap Y \in \mathscr{B}_Y$ such that $x \in B \cap Y \subseteq U$.

Therefore $\mathscr{B}_Y$ is a basis for J_Y.

Note. $A \subseteq X$ and $\mathscr{B}$ is a basis for J, then $x \in A^\circ$ if and only if there exists $B \in \mathscr{B}$ such that $x \in B \subseteq A$.

Continuous Functions

Definition: Let $(X, \mathcal{J})$ and $(Y, \mathcal{J}')$ be topological spaces and let $f:(X, \mathcal{J}) \to (Y, \mathcal{J}')$. Then f is said to be continuous at a point $x \in X$ if for each open set V containing $f(x)$ there exists an open set U containing x such that y ∈ U implies f(y) ∈ V. If f is continuous at each $x \in X$ then we say that f is a continuous function.

Theorem 13. Let $(X, \mathcal{J})$, $(Y, \mathcal{J}')$ be topological spaces. Then a function f : X → Y is continuous if and only if for each open set *V* in *Y*, $f^{-1}(V)$ is open in X.

Proof: Let f be continuous and V be an open set in Y.

Claim: $f^{-1}(V) = \{x \in X : f(x) \in V\}$ is open in X.

If $f^{-1}(V) = \phi$ then $f^{-1}(V)$ is open in X. If $f^{-1}(V) \neq \phi$, let $x \in f^{-1}(V)$. Then $f(x) \in V$. Now f is continuous at x, V is open containing $f(x)$ implies there exists an open set U such that $x \in U$ and $x' \in U$ implies $f(x') \in V$. That is f(U) ⊆ V. This implies U ⊆ f^{-1}(f(U)). Therefore x is an interior point of f^{-1}(V). Hence f^{-1}(V) is open in X.

To prove the converse, assume that f^{-1}(V) is open in X whenever V is open in Y. Now take $x \in X$ and an open set V in Y such that $f(x) \in V$. Now V is open in Y implies f^{-1}(V) is open in X. Also $f(x) \in V$ implies $x \in f^{-1}(V) = U$. That is U is an open set in X containing x such that y ∈ U implies f(y) ∈ f(f^{-1}(V)) ⊆ V. Hence f is continuous at each $x \in X$.

Theorem 14. A function f : X → Y continuous if and only if f^{-1}(A) is closed in X whenever A is closed in Y.

Proof: Assume that f : X → Y is a continuous function. Take a closed set A in Y. Since A is a closed set in Y, A^c is an open set in Y. Therefore f is a continuous function implies $f^{-1}(A^c) = [f^{-1}(A)]^c$ is an open set in X. This proves that f^{-1}(A) is a closed set in X.

To prove the converse, assume that f^{-1}(A) is closed in X whenever A is closed in Y. Take an open set V in Y. Now V is an open set in Y implies $f^{-1}(V^c) = [f^{-1}(V)]^c$ is a closed set in X. Therefore $f^{-1}(V^c)$ is a closed set in X, and hence f^{-1}(V) is an open set in X. This gives that f is a continuous function.

Example: Let $X = \mathbb{R}^w = \{(x_1, x_2, \ldots, x_n, \ldots) : x_n \in \mathbb{R}, n \in \mathbb{N}\}$ and let $\mathcal{B} = \{U_1 \times U_2 \times \cdots \times U_k \times \mathbb{R} \times \mathbb{R} \cdots : k \in \mathbb{N}$, each U_i is open in $\mathbb{R}$, $i = 1, 2, \ldots, k\}$. For $A, B \in \mathcal{B}$, $A \cap B = (U_1 \cap V_1) \times \cdots \times (U_k \cap V_k) \times \mathbb{R} \times \mathbb{R} \times \cdots \in \mathcal{B}$. Then $\mathcal{B}$ is a basis for a topology on $\mathbb{R}^w$. The topology $\mathcal{J}$ on $\mathbb{R}^w$ induced by $\mathcal{B}$ is called the product topology on $\mathbb{R}^w$. $\mathcal{B}_b = \{U_1 \times U_2 \times \cdots \times U_k \times U_{k+1} \times \cdots$ each U_k is open in $\mathbb{R}, \forall k \in \mathbb{N}\} = \{\prod_{k=1}^{\infty} U_k : U_k \textit{ is open in } \mathbb{R}\ \forall\ k\}$. Then $\mathcal{B}_b$ is also a basis for a topology on

$\mathbb{R}^w$. Let $\mathcal{J}_b$ be the topology on $\mathbb{R}^w$ induced by $\mathcal{B}_b$. This topology on $\mathbb{R}^w$ is called the box topology on $\mathbb{R}^w$.

Example: Define $f:\mathbb{R}\to(\mathbb{R}^w, \mathcal{J}_b)$ by $f(t)=(t, t, t, \ldots)$ and $U=(-1,1)\times\left(\frac{-1}{2},\frac{1}{2}\right)\times\left(\frac{-1}{3},\frac{1}{3}\right)$ ... $\times\cdots=\prod_{n=1}^{\infty}\left(\frac{-1}{n},\frac{1}{n}\right)$.

Then $U\in\mathcal{J}_b, f^{-1}(U)=\{t\in\mathbb{R}: f(t)\in U\}=\{t\in\mathbb{R}:(t, t\ldots)\in\prod_{n=1}^{\infty}\left(\frac{-1}{n},\frac{1}{n}\right)=U\}$

$=\{t\in\mathbb{R}:|t|<\frac{1}{n},\forall n\in\mathbb{N}\}=\{0\}$, and {o} is not an open set in $\mathbb{R}$. Hence f is not a continuous function. But the same $f:\mathbb{R}\to(\mathbb{R}^w, \mathcal{J})$ is a continuous function, when we consider the product topology $\mathcal{J}$ on $\mathbb{R}^w$.

Theorem 15. A function f : X → Y is continuous if and only if for every subset A of X, $f(\overline{A})\subseteq\overline{f(A)}$ (where it is understood that X, Y are topological spaces).
Proof: Now assume that f : X → Y is continuous. To prove for $A\subseteq X, f(\overline{A})\subseteq\overline{f(A)}$.

Now $\overline{f(A)}$ is a closed set in Y and f : X → Y is a continuous function implies $f^{-1}(\overline{f(A)})$ is a closed set in X. Also $A\subseteq f^{-1}(f(A))\subseteq f^{-1}(\overline{f(A)})$. That is $f^{-1}(\overline{f(A)})$ is a closed set containing A. Hence $\overline{A}\subseteq f^{-1}(\overline{f(A)})$. This gives that $f(\overline{A})\subseteq f(f^{-1}(\overline{f(A)}))\subseteq\overline{f(A)}$.
Conversely assume that for $A\subseteq X$ $f(\overline{A})\subseteq\overline{f(A)}$. Let F be a closed set in Y and A =$f^{-1}(F)$.
Now $f(\overline{A})\subseteq\overline{f(A)}$ implies $f(\overline{f^{-1}(F)})\subseteq\overline{f(f^{-1}(F))}\subseteq\overline{F}=F$. Hence $f^{-1}(f(\overline{f^{-1}(F)}))\subseteq f^{-1}(F)$.
This gives that $\overline{f^{-1}(F)}\subseteq f^{-1}(F)$. This proves that $f^{-1}(F)$ is a closed set in X whenever F is a closed set in Y. Therefore f : X → Y is a continuous function.

Remark: Intuitively what do we mean by a continuous function? In the above theorem, for any subset A of X, if a point x is closer to A then the image $f(x)$ is closer to f(A). Here $x\in\overline{A}$ means x is closer to A and hence $f(x)\in f(\overline{A})$. Now we want that $f(x)$ is closer to f(A). That is $f(x)\in\overline{f(A)}$. So a function f : X → Y is continuous if and only if for every subset A of X, x is closer to A implies $f(x)$ is closer to f(A).

Definition: Let X, Y be topological spaces. Then a function f : X → Y is said to be a homeomorphism if and only if

(i) *f* is bijective

(ii) $f: X\to Y$, $f^{-1}: Y\to X$ are continuous.

Example: (i) f: [0, 1] → [a, b] defined by $f(t)=(1-t)a+tb$ is a homeomorphism.

(ii) f: (0, 1) → (1,∞) defined by $f(t)=\frac{1}{t}$ is a homeomorphism.

(iii) $f:(0,1)\to(0,\infty)$ defined by $f(t)=\dfrac{t}{1-t}$ is a homeomorphism.

(iv) Let $X=(\mathbb{R},\mathcal{J}_s)$, $Y=(\mathbb{R},\mathcal{J}_f)$. Let $F\neq\phi$ be a closed in Y. Then F is a finite set or $F=\mathbb{R}$. In any case $f^{-1}(F)=F$ is closed in X. Hence f is continuous but the identity map $f^{-1}: Y\to X$ is not continuous.

Example: Let X = [(−1, −1),(1, −1)] be the line segment joining the points (−1, −1) and (1, −1) in $\mathbb{R}^2$ and $Y=\{(x,y)\in\mathbb{R}^2:-1\le x\le 1,\ y\ge 0,\ x^2+y^2=1\}$. Then X, Y are subspaces of the Euclidean space $\mathbb{R}^2$.

Define $f: X\to Y$ as $f((x,y))=f(x,-1)=\left(x,\sqrt{1-x^2}\right)$ then f is a homeomorphism. That is f is bijective and U is open in X if and only if $f(U)$ is open in Y.

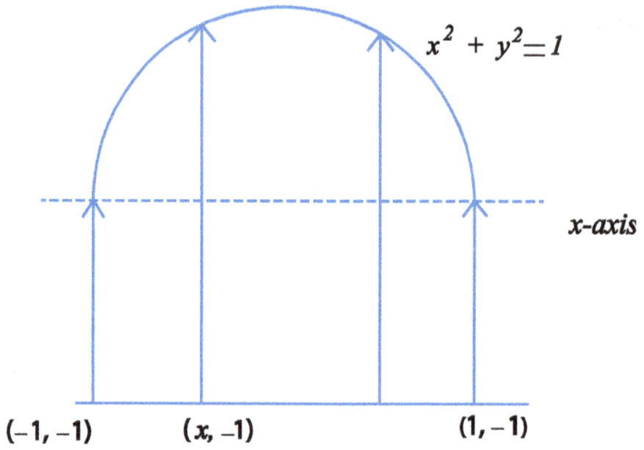

Exercise: Let A and B be two distinct points in $\mathbb{R}^2$ and γ be a curve joining A and B as shown below: That is $\gamma:[0,1]\to\mathbb{R}^2$ is a one-one continuous function.

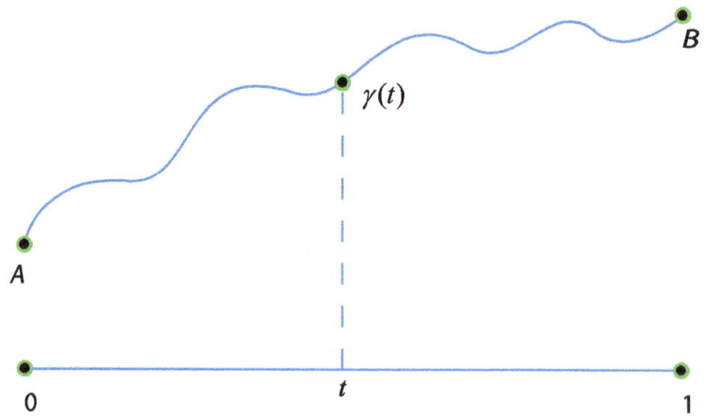

Then prove that $\gamma:[0,1]\to\{\gamma(t): t\in[0,1]\}$ is a homeomorphism. That is [0,1] and $\{\gamma(t): t\in[0,1]\}$ are equivalent topological spaces. That is, there is a homeomorphism between these two topological spaces.

Exercise: Prove that $f: X \to Y$ is a homeomorphism if and only if

(i) f is bijective

(ii) $f(\bar{A}) = \overline{f(A)}$, for $A \subseteq X$

Connectedness

In mathematics, connectedness is used to refer to various properties meaning, in some sense, "all one piece". When a mathematical object has such a property, we say it is connected; otherwise it is disconnected. When a disconnected object can be split naturally into connected pieces, each piece is usually called a *component* (or *connected component*).

Connectedness in Topology

A topological space is said to be *connected* if it is not the union of two disjoint non-empty open sets. A set is open if it contains no point lying on its boundary; thus, in an informal, intuitive sense, the fact that a space can be partitioned into disjoint open sets suggests that the boundary between the two sets is not part of the space, and thus splits it into two separate pieces.

Other Notions of Connectedness

Fields of mathematics are typically concerned with special kinds of objects. Often such an object is said to be *connected* if, when it is considered as a topological space, it is a connected space. Thus, manifolds, Lie groups, and graphs are all called *connected* if they are connected as topological spaces, and their components are the topological components. Sometimes it is convenient to restate the definition of connectedness in such fields. For example, a graph is said to be *connected* if each pair of vertices in the graph is joined by a path. This definition is equivalent to the topological one, as applied to graphs, but it is easier to deal with in the context of graph theory. Graph theory also offers a context-free measure of connectedness, called the clustering coefficient.

Other fields of mathematics are concerned with objects that are rarely considered as topological spaces. Nonetheless, definitions of *connectedness* often reflect the topological meaning in some way. For example, in category theory, a category is said to be *connected* if each pair of objects in it is joined by a sequence of morphisms. Thus, a category is connected if it is, intuitively, all one piece.

There may be different notions of *connectedness* that are intuitively similar, but different as formally defined concepts. We might wish to call a topological space *connected* if each pair of points in it is joined by a path. However this concept turns out to be different from standard topological connectedness; in particular, there are connected topological spaces for which this property does not hold. Because of this, different terminology is used; spaces with this property are said to be *path connected*. While not all connected spaces are path connected, all path connected spaces are connected.

Terms involving *connected* are also used for properties that are related to, but clearly different from, connectedness. For example, a path-connected topological space is *simply connected* if each loop (path from a point to itself) in it is contractible; that is, intuitively, if there is essentially only one way to get from any point to any other point. Thus, a sphere and a disk are each simply connected, while a torus is not. As another example, a directed graph is *strongly connected* if each ordered pair of vertices is joined by a directed path (that is, one that "follows the arrows").

Other concepts express the way in which an object is *not* connected. For example, a topological space is *totally disconnected* if each of its components is a single point.

Connectivity

Properties and parameters based on the idea of connectedness often involve the word *connectivity*. For example, in graph theory, a connected graph is one from which we must remove at least one vertex to create a disconnected graph. In recognition of this, such graphs are also said to be *1-connected*. Similarly, a graph is *2-connected* if we must remove at least two vertices from it, to create a disconnected graph. A *3-connected* graph requires the removal of at least three vertices, and so on. The *connectivity* of a graph is the minimum number of vertices that must be removed, to disconnect it. Equivalently, the connectivity of a graph is the greatest integer k for which the graph is k-connected.

While terminology varies, noun forms of connectedness-related properties often include the term *connectivity*. Thus, when discussing simply connected topological spaces, it is far more common to speak of *simple connectivity* than *simple connectedness*. On the other hand, in fields without a formally defined notion of *connectivity*, the word may be used as a synonym for *connectedness*.

Another example of connectivity can be found in regular tilings. Here, the connectivity describes the number of neighbors accessible from a single tile:

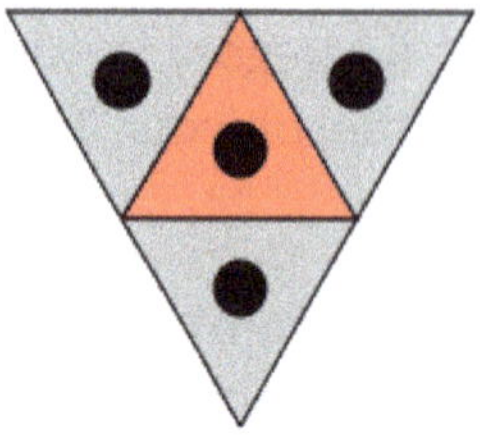

3-connectivity in a triangular tiling,

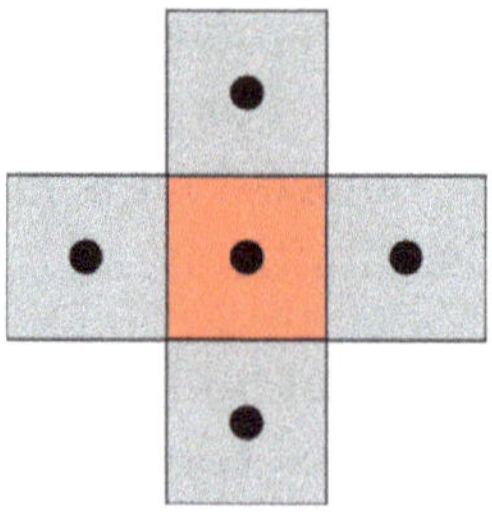

4-connectivity in a square tiling,

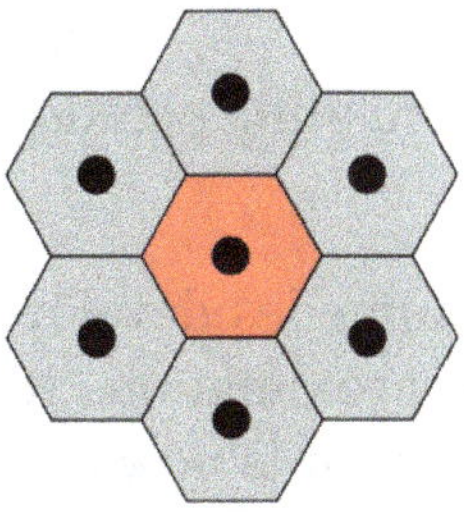

6-connectivity in a hexagonal tiling,

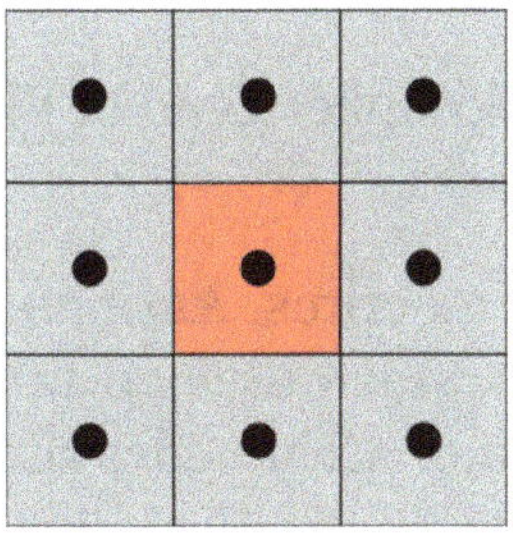

8-connectivity in a square tiling (note that distance equity is not kept)

Limit of a Sequence

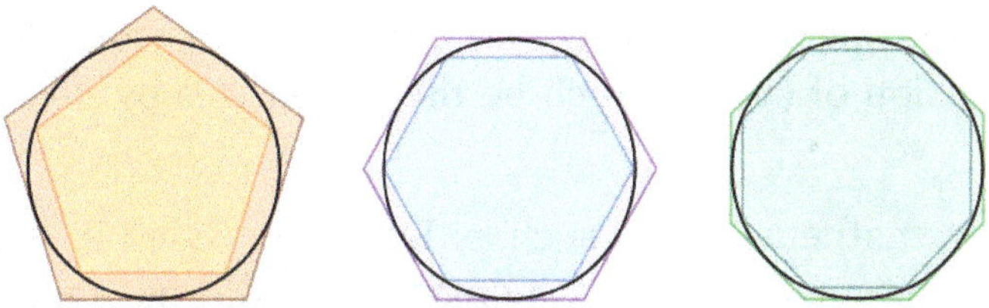

The sequence given by the perimeters of regular n-sided polygons that circumscribe the unit circle has a limit equal to the perimeter of the circle, i.e. $2\pi r$. The corresponding sequence for inscribed polygons has the same limit.

n	n sin(1/n)
1	0.841471
2	0.958851
...	
10	0.998334
...	
100	0.999983
As the positive integer n becomes larger and larger, the value $n \cdot \sin\left(\frac{1}{n}\right)$ becomes arbitrarily close to 1. We say that "the limit of the sequence $n \cdot \sin\left(\frac{1}{n}\right)$ equals 1."	

In mathematics, the limit of a sequence is the value that the terms of a sequence "tend to". If such a limit exists, the sequence is called convergent. A sequence which does not converge is said to be divergent. The limit of a sequence is said to be the fundamental notion on which the whole of analysis ultimately rests.

Limits can be defined in any metric or topological space, but are usually first encountered in the real numbers.

History

The Greek philosopher Zeno of Elea is famous for formulating paradoxes that involve limiting processes.

Leucippus, Democritus, Antiphon, Eudoxus and Archimedes developed the method of exhaustion, which uses an infinite sequence of approximations to determine an area or a volume. Archimedes succeeded in summing what is now called a geometric series.

Newton dealt with series in his works on *Analysis with infinite series* (written in 1669, circulated in manuscript, published in 1711), *Method of fluxions and infinite series* (written in 1671, published in English translation in 1736, Latin original published much later) and *Tractatus de Quadratura Curvarum* (written in 1693, published in 1704 as an Appendix to his *Optiks*). In the latter work, Newton considers the binomial expansion of $(x+o)^n$ which he then linearizes by *taking limits* (letting $o \to 0$).

In the 18th century, mathematicians such as Euler succeeded in summing some *divergent* series by stopping at the right moment; they did not much care whether a limit existed, as long as it could be calculated. At the end of the century, Lagrange in his *Théorie des fonctions analytiques* (1797) opined that the lack of rigour precluded further development in calculus. Gauss in his etude of hypergeometric series (1813) for the first time rigorously investigated under which conditions a series converged to a limit.

The modern definition of a limit (for any ε there exists an index N so that ...) was given by Bernhard Bolzano (*Der binomische Lehrsatz*, Prague 1816, little noticed at the time) and by Karl Weierstrass in the 1870s.

Real Numbers

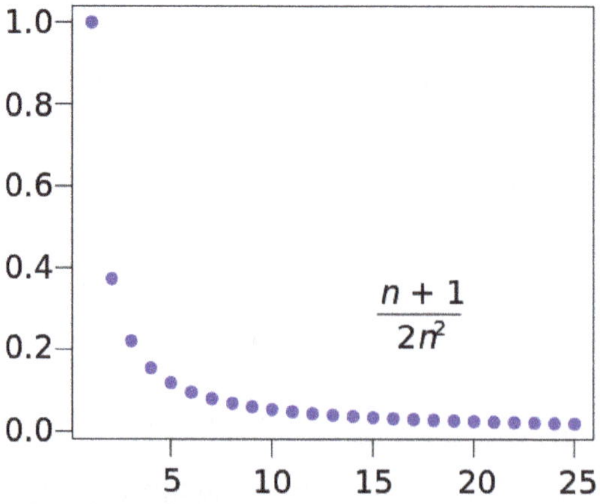

The plot of a convergent sequence $\{a_n\}$ is shown in blue.
Visually we can see that the sequence is converging to the limit 0 as n increases.

In the real numbers, a number L is the limit of the sequence (x_n) if the numbers in the sequence become closer and closer to L and not to any other number.

Examples

- If $x_n = c$ for constant c, then $x_n \to c$.
- If $x_n = \frac{1}{n}$, then $x_n \to 0$.
- If $x_n = 1/n$ when n is even, and $x_n = \frac{1}{n^2}$ when n is odd, then $x_n \to 0$. (The fact that $x_{n+1} > x_n$ whenever n is odd is irrelevant.)
- Given any real number, one may easily construct a sequence that converges to that number by taking decimal approximations. For example, the sequence $0.3, 0.33, 0.333, 0.3333, \ldots$ converges to $1/3$. Note that the decimal representation $0.3333\ldots$ is the *limit* of the previous sequence, defined by

$$0.3333\ldots \triangleq \lim_{n\to\infty} \sum_{i=1}^{n} \frac{3}{10^i}.$$

- Finding the limit of a sequence is not always obvious. Two examples are $\lim_{n\to\infty}\left(1+\frac{1}{n}\right)^n$ (the limit of which is the number e) and the Arithmetic–geometric mean. The squeeze theorem is often useful in such cases.

Formal Definition

We call x the limit of the sequence (x_n) if the following condition holds:

- For each real number $\epsilon > 0$, there exists a natural number N such that, for every natural number $n \geq N$, we have $|x_n - x| < \epsilon$.

In other words, for every measure of closeness ϵ, the sequence's terms are eventually that close to the limit. The sequence (x_n) is said to converge to or tend to the limit x, written $x_n \to x$ or $\lim_{n\to\infty} x_n = x$.

Symbolically, this is:

- $\forall \epsilon > 0 (\exists N \in \mathbb{N} (\forall n \in \mathbb{N} (n \geq N \Rightarrow |x_n - x| < \epsilon)))$.

If a sequence converges to some limit, then it is convergent; otherwise it is divergent.

Illustration

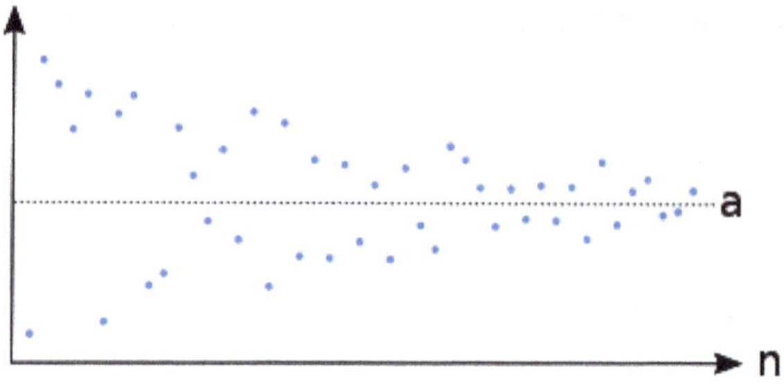

Example of a sequence which converges to the limit a.

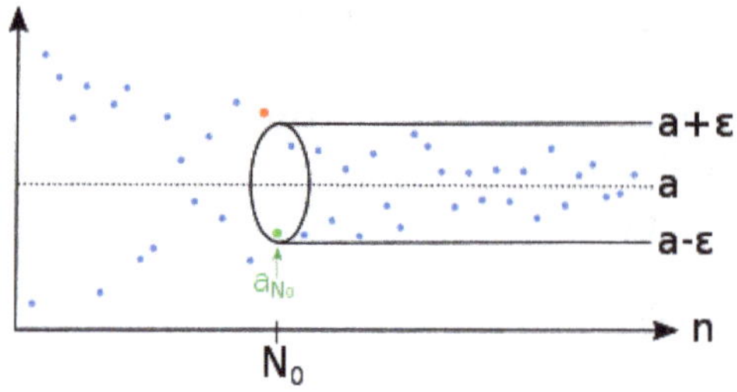

Regardless which $\varepsilon > 0$ we have, there is an index N_0, so that the sequence lies afterwards completely in the epsilon tube $(a-\varepsilon, a+\varepsilon)$.

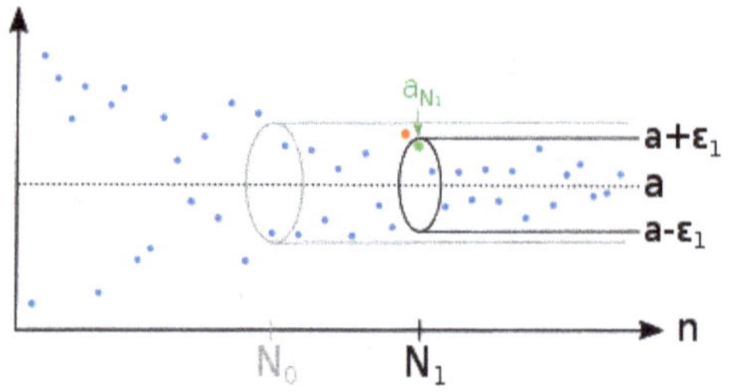

There is also for a smaller $\epsilon_1 > 0$ an index N_1, so that the sequence is afterwards inside the epsilon tube $(a-\varepsilon_1, a+\varepsilon_1)$.

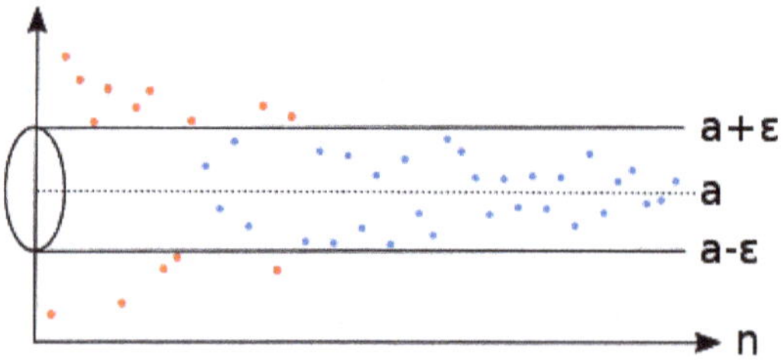

For each $\varepsilon > 0$ there are only finitely many sequence members outside the epsilon tube.

Properties

Limits of sequences behave well with respect to the usual arithmetic operations. If $a_n \to a$ and $b_n \to b$, then $a_n + b_n \to a + b$, $a_n \cdot b_n \to ab$ and, if neither b nor any b_n is zero, $\frac{a_n}{b_n} \to \frac{a}{b}$.

For any continuous function *f*, if $x_n \to x$ then $f(x_n) \to f(x)$. In fact, any real-valued function *f* is continuous if and only if it preserves the limits of sequences (though this is not necessarily true when using more general notions of continuity).

Some other important properties of limits of real sequences include the following (provided, in each equation below, that the limits on the right exist).

- The limit of a sequence is unique.
- $\lim\limits_{n\to\infty}(a_n \pm b_n) = \lim\limits_{n\to\infty} a_n \pm \lim\limits_{n\to\infty} b_n$
- $\lim\limits_{n\to\infty} ca_n = c \cdot \lim\limits_{n\to\infty} a_n$
- $\lim\limits_{n\to\infty}(a_n \cdot b_n) = (\lim\limits_{n\to\infty} a_n) \cdot (\lim\limits_{n\to\infty} b_n)$
- $\lim\limits_{n\to\infty}\left(\frac{a_n}{b_n}\right) = \frac{\lim\limits_{n\to\infty} a_n}{\lim\limits_{n\to\infty} b_n}$ provided $\lim\limits_{n\to\infty} b_n \neq 0$
- $\lim\limits_{n\to\infty} a_n^p = \left[\lim\limits_{n\to\infty} a_n\right]^p$
- If $a_n \leq b_n$ for all n greater than some N, then $\lim\limits_{n\to\infty} a_n \leq \lim\limits_{n\to\infty} b_n$
- (Squeeze theorem) If $a_n \leq c_n \leq b_n$ for all $n > N$, and $\lim\limits_{n\to\infty} a_n = \lim\limits_{n\to\infty} b_n = L$, then $\lim\limits_{n\to\infty} c_n = L$.
- If a sequence is bounded and monotonic then it is convergent.
- A sequence is convergent if and only if every subsequence is convergent.

These properties are extensively used to prove limits without the need to directly use the cumbersome formal definition. Once proven that $\frac{1}{n} \to 0$ it becomes easy to show that $\frac{a}{b+\frac{c}{n}} \to \frac{a}{b}$, ($b \neq 0$), using the properties above.

Infinite Limits

A sequence (x_n) is said to tend to infinity, written $x_n \to \infty$ or $\lim\limits_{n\to\infty} x_n = \infty$ if, for every *K*, there is an *N* such that, for every $n \geq N$, $x_n > K$; that is, the sequence terms are eventually larger than any fixed *K*. Similarly, $x_n \to -\infty$ if, for every *K*, there is an *N* such that, for every $n \geq N$, $x_n < K$. If a sequence tends to infinity, or to minus infinity, then it is divergent (however, a divergent sequence need not tend to plus or minus infinity: take for example $x_n = (-1)^n$).

Metric Spaces

Definition

A point *x* of the metric space (*X*, *d*) is the limit of the sequence (x_n) if, for all ε > 0, there is an N such that, for every $n \geq N$, $d(x_n, x) < \epsilon$. This coincides with the definition given for real numbers when $X = \mathbb{R}$ and $d(x, y) = |x - y|$.

Properties

For any continuous function *f*, if $x_n \to x$ then $f(x_n) \to f(x)$. In fact, a function *f* is continuous if and only if it preserves the limits of sequences.

Limits of sequences are unique when they exist, as distinct points are separated by some positive distance, so for ϵ less than half this distance, sequence terms cannot be within a distance ϵ of both points.

Topological Spaces

Definition

A point x of the topological space (X, τ) is the limit of the sequence (x_n) if, for every neighbourhood U of x, there is an N such that, for every $n \geq N$, $x_n \in U$. This coincides with the definition given for metric spaces if (X,d) is a metric space and τ is the topology generated by d.

The limit of a sequence of points $(x_n : n \in \mathbb{N})$ in a topological space T is a special case of the limit of a function: the domain is $\mathbb{N}$ in the space $\mathbb{N} \cup \{+\infty\}$ with the induced topology of the affinely extended real number system, the range is T, and the function argument n tends to $+\infty$, which in this space is a limit point of $\mathbb{N}$.

Properties

If X is a Hausdorff space then limits of sequences are unique where they exist. Note that this need not be the case in general; in particular, if two points x and y are topologically indistinguishable, any sequence that converges to x must converge to y and vice versa.

Cauchy Sequences

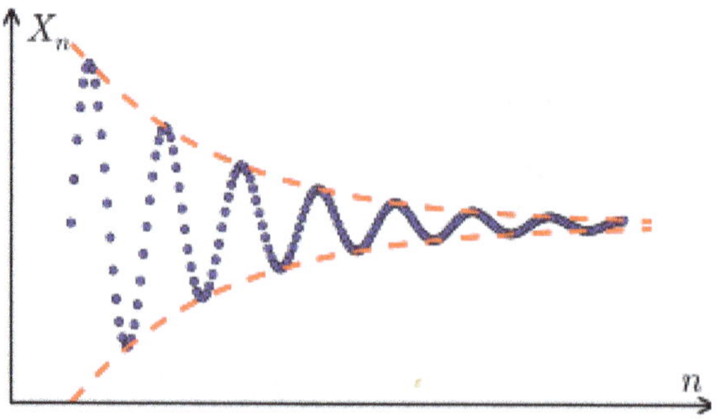

The plot of a Cauchy sequence (x_n), shown in blue, as x_n versus n. Visually, we see that the sequence appears to be converging to a limit point as the terms in the sequence become closer together as n increases. In the real numbers every Cauchy sequence converges to some limit.

A Cauchy sequence is a sequence whose terms ultimately become arbitrarily close together, after sufficiently many initial terms have been discarded. The notion of a Cauchy sequence is important in the study of sequences in metric spaces, and, in particular, in real analysis. One particularly important result in real analysis is the *Cauchy criterion for convergence of sequences*: A sequence of real numbers is convergent if and only if it is a Cauchy sequence. This remains true in other complete metric spaces.

Definition in Hyperreal Numbers

The definition of the limit using the hyperreal numbers formalizes the intuition that for a "very large" value of the index, the corresponding term is "very close" to the limit. More precisely, a real sequence (x_n) tends to L if for every infinite hypernatural H, the term x_H is infinitely close to L, i.e., the difference $x_H - L$ is infinitesimal. Equivalently, L is the standard part of x_H

$$L = st(x_H).$$

Thus, the limit can be defined by the formula

$$\lim_{n\to\infty} x_n = st(x_H),$$

where the limit exists if and only if the righthand side is independent of the choice of an infinite H.

Manifold

In mathematics, a manifold is a topological space that locally resembles Euclidean space near each point. More precisely, each point of an n-dimensional manifold has a neighbourhood that is homeomorphic to the Euclidean space of dimension n. In this more precise terminology, a manifold is referred to as an n-manifold.

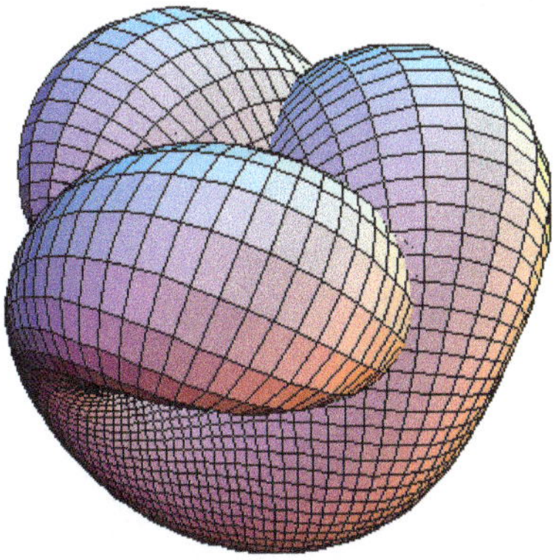

The real projective plane is a two-dimensional manifold that cannot be realized in three dimensions without self-intersection, shown here as Boy's surface.

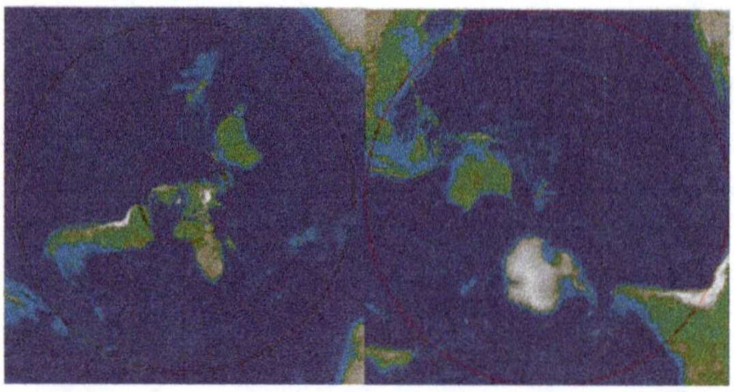

The surface of the Earth requires (at least) two charts to include every point. Here the globe is decomposed into charts around the North and South Poles.

One-dimensional manifolds include lines and circles, but not figure eights (because they have *crossing points* that are not locally homeomorphic to Euclidean 1-space). Two-dimensional manifolds are also called surfaces. Examples include the plane, the sphere, and the torus, which can all be embedded (formed without self-intersections) in three dimensional real space, but also the Klein bottle and real projective plane, which will always self-intersect when immersed in three-dimensional real space.

Although a manifold locally resembles Euclidean space, meaning that every point has a neighborhood homeomorphic to an open subset of Euclidean space, globally it may not: manifolds in general are not homeomorphic to Euclidean space. For example, the surface of the sphere is not homeomorphic to a Euclidean space, because (among other properties) it has the global topological property of compactness that Euclidean space lacks, but in a region it can be charted by means of map projections of the region into the Euclidean plane (in the context of manifolds they are called *charts*). When a region appears in two neighbouring charts, the two representations do not coincide exactly and a transformation is needed to pass from one to the other, called a *transition map*.

The concept of a manifold is central to many parts of geometry and modern mathematical physics because it allows complicated structures to be described and understood in terms of the simpler local topological properties of Euclidean space. Manifolds naturally arise as solution sets of systems of equations and as graphs of functions.

Manifolds can be equipped with additional structure. One important class of manifolds is the class of differentiable manifolds; this differentiable structure allows calculus to be done on manifolds. A Riemannian metric on a manifold allows distances and angles to be measured. Symplectic manifolds serve as the phase spaces in the Hamiltonian formalism of classical mechanics, while four-dimensional Lorentzian manifolds model spacetime in general relativity.

Motivating Examples

A surface is a two dimensional manifold, meaning that it locally resembles the Euclidean plane near each point. For example, the surface of a globe can be described by a collection of maps (called charts), which together form an atlas of the globe. Although no individual map is sufficient to cover the entire surface of the globe, any place in the globe will be in at least one of the charts.

Many places will appear in more than one chart. For example, a map of North America will likely include parts of South America and the Arctic circle. These regions of the globe will be described in full in separate charts, which in turn will contain parts of North America. There is a relation between adjacent charts, called a *transition map* that allows them to be consistently patched together to cover the whole of the globe.

Describing the coordinate charts on surfaces explicitly requires knowledge of functions of two variables, because these patching functions must map a region in the plane to another region of the plane. However, one-dimensional examples of manifolds (or curves) can be described with functions of a single variable only.

Circle

After a line, the circle is the simplest example of a topological manifold. Topology ignores bending, so a small piece of a circle is treated exactly the same as a small piece of a line. Consider, for instance, the top part of the unit circle, $x^2 + y^2 = 1$, where the y-coordinate is positive (indicated by the yellow circular arc in *Figure*). Any point of this arc can be uniquely described by its x-coordinate. So, projection onto the first coordinate is a continuous, and invertible, mapping from the upper arc to the open interval $(-1, 1)$:

$$\chi_{top}(x, y) = x.$$

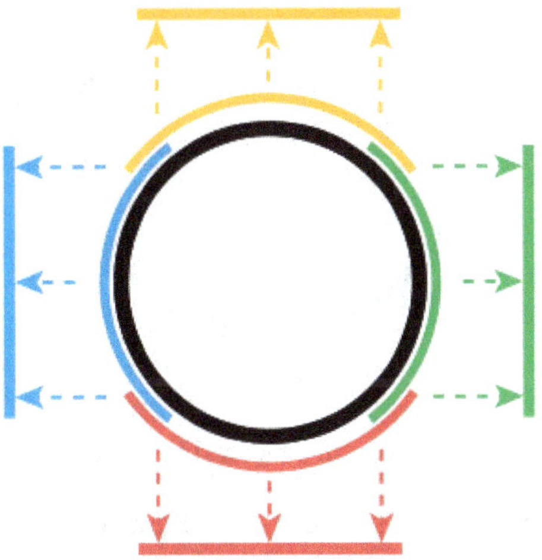

The four charts each map part of the circle to an open interval, and together cover the whole circle.

Such functions along with the open regions they map are called *charts*. Similarly, there are charts for the bottom (red), left (blue), and right (green) parts of the circle:

$$\chi_{\text{bottom}}(x, y) = x$$
$$\chi_{\text{left}}(x, y) = y$$
$$\chi_{\text{right}}(x, y) = y.$$

Together, these parts cover the whole circle and the four charts form an atlas for the circle.

The top and right charts, χ_{top} and χ_{right} respectively, overlap in their domain: their intersection lies in the quarter of the circle where both the *x*- and the *y*-coordinates are positive. Each map this part into the interval (0, 1), though differently. Thus a function $T:(0,1) \to (0,1) = \chi_{right} \circ \chi_{top}^{-1}$ can be constructed, which takes values from the co-domain of χ_{top} back to the circle using the inverse, followed by the χ_{right} back to the interval. Let *a* be any number in (0, 1), then:

$$\begin{aligned} T(a) &= \chi_{\text{right}}\left(\chi_{\text{top}}^{-1}[a]\right) \\ &= \chi_{\text{right}}\left(a, \sqrt{1-a^2}\right) \\ &= \sqrt{1-a^2} \end{aligned}$$

Such a function is called a *transition map.*

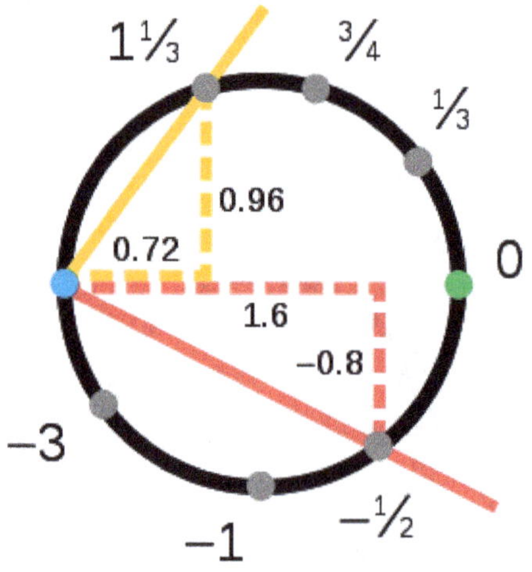

A circle manifold chart based on slope, covering all but one point of the circle.

The top, bottom, left, and right charts show that the circle is a manifold, but they do not form the only possible atlas. Charts need not be geometric projections, and the number of charts is a matter of choice. Consider the charts

$$\chi_{\text{minus}}(x, y) = s = \frac{y}{1+x}$$

and

$$\chi_{\text{plus}}(x, y) = t = \frac{y}{1-x}$$

Here s is the slope of the line through the point at coordinates (x,y) and the fixed pivot point $(-1, 0)$; t follows similarly, but with pivot point $(+1, 0)$. The inverse mapping from s to (x, y) is given by

$$x = \frac{1-s^2}{1+s^2}$$
$$y = \frac{2s}{1+s^2}$$

It can easily be confirmed that $x^2 + y^2 = 1$ for all values of the slope s. These two charts provide a second atlas for the circle, with

$$t = \frac{1}{s}$$

Each chart omits a single point, either $(-1, 0)$ for s or $(+1, 0)$ for t, so neither chart alone is sufficient to cover the whole circle. It can be proved that it is not possible to cover the full circle with a single chart. For example, although it is possible to construct a circle from a single line interval by overlapping and "gluing" the ends, this does not produce a chart; a portion of the circle will be mapped to both ends at once, losing invertibility.

Sphere

The sphere is an example of a surface. The unit sphere of implicit equation

$$x^2 + y^2 + z^2 - 1 = 0$$

may be covered by an atlas of six charts: the plane $z = 0$ divides the sphere into two half spheres ($z > 0$ and $z < 0$), which may both be mapped on the disc $x^2 + y^2 < 1$ by the projection on the xy plane of coordinates. This provides two charts; the four other charts are provided by a similar construction with the two other coordinate planes.

As for the circle, one may define one chart that covers the whole sphere excluding one point. Thus two charts are sufficient, but the sphere cannot be covered by a single chart.

This example is historically significant, as it has motivated the terminology; it became apparent that the whole surface of the Earth cannot have a plane representation consisting of a single map (also called "chart"), and therefore one needs atlases for covering the whole Earth surface.

Enriched Circle

Viewed using calculus, the circle transition function T is simply a function between open intervals, which gives a meaning to the statement that T is differentiable. The transition map T, and all the others, are differentiable on (0, 1); therefore, with this atlas the circle is a *differentiable manifold*. It is also *smooth* and *analytic* because the transition functions have these properties as well.

Other circle properties allow it to meet the requirements of more specialized types of manifold. For example, the circle has a notion of distance between two points, the arc-length between the points; hence it is a *Riemannian manifold*.

Other Curves

Four manifolds from algebraic curves: ■ circles, ■ parabola, ■ hyperbola, ■ cubic.

Manifolds need not be connected (all in "one piece"); an example is a pair of separate circles.

Manifolds need not be closed; thus a line segment without its end points is a manifold. And they are never countable, unless the dimension of the manifold is 0. Putting these freedoms together, other examples of manifolds are a parabola, a hyperbola (two open, infinite pieces), and the locus of points on a cubic curve $y^2 = x^3 - x$ (a closed loop piece and an open, infinite piece).

However, excluded are examples like two touching circles that share a point to form a figure; at the shared point a satisfactory chart cannot be created. Even with the bending allowed by topology, the vicinity of the shared point looks like a "+", not a line. A "+" is not homeomorphic to a closed interval (line segment), since deleting the center point from the "+" gives a space with four components (i.e. pieces), whereas deleting a point from a closed interval gives a space with at most two pieces; topological operations always preserve the number of pieces.

Mathematical Definition

Informally, a manifold is a space that is "modeled on" Euclidean space.

There are many different kinds of manifolds, depending on the context. In geometry and topology, all manifolds are topological manifolds, possibly with additional structure, such as a differentiable structure. A manifold can be constructed by giving a collection of coordinate charts, that is a covering by open sets with homeomorphisms to a Euclidean space, and patching functions: homeomorphisms from one region of Euclidean space to another region if they correspond to the same part of the manifold in two different coordinate charts. A manifold can be given additional structure if the patching functions satisfy axioms beyond continuity. For instance, differentiable manifolds have homeomorphisms on overlapping neighborhoods diffeomorphic with each other, so that the manifold has a well-defined set of functions which are differentiable in each neighborhood, and so differentiable on the manifold as a whole.

Formally, a (topological) manifold is a second countable Hausdorff space that is locally homeomorphic to Euclidean space.

Second countable and *Hausdorff* are point-set conditions; *second countable* excludes spaces which are in some sense 'too large' such as the long line, while *Hausdorff* excludes spaces such as "the line with two origins" (these generalizations of manifolds are discussed in non-Hausdorff manifolds).

Locally homeomorphic to Euclidean space means that every point has a neighborhood homeomorphic to an open Euclidean n-ball,

$$\mathrm{B}^n = \{(x_1, x_2, \ldots, x_n) \in \mathbb{R}^n \mid x_1^2 + x_2^2 + \cdots + x_n^2 < 1\}.$$

More precisely, locally homeomorphic here means that each point m in the manifold M has an open neighborhood homeomorphic to an open *neighborhood* in Euclidean space, not to the unit ball specifically. However, given such a homeomorphism, the pre-image of an ϵ -ball gives a homeomorphism between the unit ball and a smaller neighborhood of m, so this is no loss of generality. For topological or differentiable manifolds, one can also ask that every point have a neighborhood homeomorphic to all of Euclidean space (as this is diffeomorphic to the unit ball), but this cannot be done for complex manifolds, as the complex unit ball is not holomorphic to complex space.

Generally manifolds are taken to have a fixed dimension (the space must be locally homeomorphic to a fixed n-ball), and such a space is called an n-manifold; however, some authors admit manifolds where different points can have different dimensions. If a manifold has a fixed dimension, it is called a pure manifold. For example, the sphere has a constant dimension of 2 and is therefore a pure manifold whereas the disjoint union of a sphere and a line in three-dimensional space is *not* a pure manifold. Since dimension is a local invariant (i.e. the map sending each point to the dimension of its neighbourhood over which a chart is defined, is locally constant), each connected component has a fixed dimension.

Scheme-theoretically, a manifold is a locally ringed space, whose structure sheaf is locally isomorphic to the sheaf of continuous (or differentiable, or complex-analytic, etc.) functions on Euclidean space. This definition is mostly used when discussing analytic manifolds in algebraic geometry.

Charts, Atlases, and Transition Maps

The spherical Earth is navigated using flat maps or charts, collected in an atlas. Similarly, a differentiable manifold can be described using mathematical maps, called *coordinate charts*, collected in a mathematical *atlas*. It is not generally possible to describe a manifold with just one chart, because the global structure of the manifold is different from the simple structure of the charts. For example, no single flat map can represent the entire Earth without separation of adjacent features across the map's boundaries or duplication of coverage. When a manifold is constructed from multiple overlapping charts, the regions where they overlap carry information essential to understanding the global structure.

Charts

A coordinate map, a coordinate chart, or simply a chart, of a manifold is an invertible map between a subset of the manifold and a simple space such that both the map and its inverse preserve the desired structure. For a topological manifold, the simple space is a subset of some Euclidean space R^n and interest focuses on the topological structure. This structure is preserved by homeomorphisms, invertible maps that are continuous in both directions.

In the case of a differentiable manifold, a set of charts called an atlas allows us to do calculus on manifolds. Polar coordinates, for example, form a chart for the plane R^2 minus the positive x-axis and the origin. Another example of a chart is the map χ_{top} mentioned in the section above, a chart for the circle.

Atlases

The description of most manifolds requires more than one chart (a single chart is adequate for only the simplest manifolds). A specific collection of charts which covers a manifold is called an atlas. An atlas is not unique as all manifolds can be covered multiple ways using different combinations of charts. Two atlases are said to be equivalent if their union is also an atlas.

The atlas containing all possible charts consistent with a given atlas is called the maximal atlas (i.e. an equivalence class containing that given atlas). Unlike an ordinary atlas, the maximal atlas of a given manifold is unique. Though it is useful for definitions, it is an abstract object and not used directly (e.g. in calculations).

Transition Maps

Charts in an atlas may overlap and a single point of a manifold may be represented in several charts. If two charts overlap, parts of them represent the same region of the manifold, just as a map of Europe and a map of Asia may both contain Moscow. Given two overlapping charts, a transition function can be defined which goes from an open ball in R^n to the manifold and then back to another (or perhaps the same) open ball in R^n. The resultant map, like the map T in the circle example above, is called a change of coordinates, a coordinate transformation, a transition function, or a transition map.

Additional Structure

An atlas can also be used to define additional structure on the manifold. The structure is first defined on each chart separately. If all the transition maps are compatible with this structure, the structure transfers to the manifold.

This is the standard way differentiable manifolds are defined. If the transition functions of an atlas for a topological manifold preserve the natural differential structure of R^n (that is, if they are diffeomorphisms), the differential structure transfers to the manifold and turns it into a differentiable manifold. Complex manifolds are introduced in an analogous way by requiring that the transition functions of an atlas are holomorphic functions. For symplectic manifolds, the transition functions must be symplectomorphisms.

The structure on the manifold depends on the atlas, but sometimes different atlases can be said to give rise to the same structure. Such atlases are called compatible.

These notions are made precise in general through the use of pseudogroups.

Manifold with Boundary

A manifold with boundary is a manifold with an edge. For example, a sheet of paper is a 2-manifold with a 1-dimensional boundary. The boundary of an n-manifold with boundary is an $(n - 1)$-manifold. A disk (circle plus interior) is a 2-manifold with boundary. Its boundary is a circle, a 1-manifold. A square with interior is also a 2-manifold with boundary. A ball (sphere plus interior) is a 3-manifold with boundary. Its boundary is a sphere, a 2-manifold.

In technical language, a manifold with boundary is a space containing both interior points and boundary points. Every interior point has a neighborhood homeomorphic to the open n-ball $\{(x_1, x_2, ..., x_n) \mid \Sigma x_i^2 < 1\}$. Every boundary point has a neighborhood homeomorphic to the "half" n-ball $\{(x_1, x_2, ..., x_n) \mid \Sigma x_i^2 < 1 \text{ and } x_1 \geq 0\}$. The homeomorphism must send each boundary point to a point with $x_1 = 0$.

Boundary and Interior

Let M be a manifold with boundary. The interior of M, denoted Int M, is the set of points in M which have neighborhoods homeomorphic to an open subset of R^n. The boundary of M, denoted ∂M, is the complement of Int M in M. The boundary points can be characterized as those points which land on the boundary hyperplane ($x_n = 0$) of R^n_+ under some coordinate chart.

If M is a manifold with boundary of dimension n, then Int M is a manifold (without boundary) of dimension n and ∂M is a manifold (without boundary) of dimension $n - 1$.

Construction

A single manifold can be constructed in different ways, each stressing a different aspect of the manifold, thereby leading to a slightly different viewpoint.

Charts

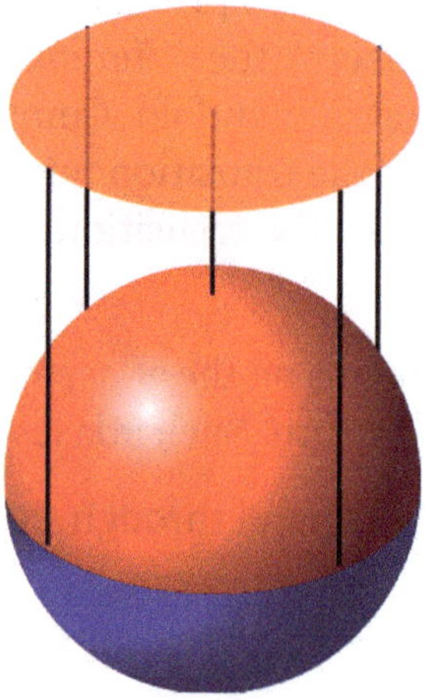

The chart maps the part of the sphere with positive z coordinate to a disc.

Perhaps the simplest way to construct a manifold is the one used in the example above of the circle. First, a subset of R^2 is identified, and then an atlas covering this subset is constructed. The concept of *manifold* grew historically from constructions like this. Here is another example, applying this method to the construction of a sphere:

Sphere with Charts

A sphere can be treated in almost the same way as the circle. In mathematics a sphere is just the surface (not the solid interior), which can be defined as a subset of R^3:

$$S = \{(x, y, z) \in R^3 \mid x^2 + y^2 + z^2 = 1\}.$$

The sphere is two-dimensional, so each chart will map part of the sphere to an open subset of R^2. Consider the northern hemisphere, which is the part with positive z coordinate (coloured red in the picture on the right). The function χ defined by

$$\chi(x,y,z)=(x,y),$$

maps the northern hemisphere to the open unit disc by projecting it on the (x, y) plane. A similar chart exists for the southern hemisphere. Together with two charts projecting on the (x, z) plane and two charts projecting on the (y, z) plane, an atlas of six charts is obtained which covers the entire sphere.

This can be easily generalized to higher-dimensional spheres.

Patchwork

A manifold can be constructed by gluing together pieces in a consistent manner, making them into overlapping charts. This construction is possible for any manifold and hence it is often used as a characterisation, especially for differentiable and Riemannian manifolds. It focuses on an atlas, as the patches naturally provide charts, and since there is no exterior space involved it leads to an intrinsic view of the manifold.

The manifold is constructed by specifying an atlas, which is itself defined by transition maps. A point of the manifold is therefore an equivalence class of points which are mapped to each other by transition maps. Charts map equivalence classes to points of a single patch. There are usually strong demands on the consistency of the transition maps. For topological manifolds they are required to be homeomorphisms; if they are also diffeomorphisms, the resulting manifold is a differentiable manifold.

This can be illustrated with the transition map $t = 1/s$ from the second half of the circle example. Start with two copies of the line. Use the coordinate s for the first copy, and t for the second copy. Now, glue both copies together by identifying the point t on the second copy with the point $s = 1/t$ on the first copy (the points $t = 0$ and $s = 0$ are not identified with any point on the first and second copy, respectively). This gives a circle.

Intrinsic and Extrinsic View

The first construction and this construction are very similar, but they represent rather different points of view. In the first construction, the manifold is seen as embedded in some Euclidean space. This is the *extrinsic view*. When a manifold is viewed in this way, it is easy to use intuition from Euclidean spaces to define additional structure. For example, in a Euclidean space it is always clear whether a vector at some point is tangential or normal to some surface through that point.

The patchwork construction does not use any embedding, but simply views the manifold as a topological space by itself. This abstract point of view is called the *intrinsic view*. It can make it harder to imagine what a tangent vector might be, and there is no intrinsic notion of a normal bundle, but instead there is an intrinsic stable normal bundle.

n-Sphere as a Patchwork

The n-sphere S^n is a generalisation of the idea of a circle (1-sphere) and sphere (2-sphere) to higher dimensions. An n-sphere S^n can be constructed by gluing together two copies of R^n. The transition map between them is defined as

$$R^n \setminus \{0\} \to R^n \setminus \{0\} : x \mapsto x / \| x \|^2 .$$

This function is its own inverse and thus can be used in both directions. As the transition map is a smooth function, this atlas defines a smooth manifold. In the case $n = 1$, the example simplifies to the circle example given earlier.

Identifying Points of a Manifold

It is possible to define different points of a manifold to be same. This can be visualized as gluing these points together in a single point, forming a quotient space. There is, however, no reason to expect such quotient spaces to be manifolds. Among the possible quotient spaces that are not necessarily manifolds, orbifolds and CW complexes are considered to be relatively well-behaved. An example of a quotient space of a manifold that is also a manifold is the real projective space identified as a quotient space of the corresponding sphere.

One method of identifying points (gluing them together) is through a right (or left) action of a group, which acts on the manifold. Two points are identified if one is moved onto the other by some group element. If M is the manifold and G is the group, the resulting quotient space is denoted by M / G (or $G \setminus M$).

Manifolds which can be constructed by identifying points include tori and real projective spaces (starting with a plane and a sphere, respectively).

Gluing along Boundaries

Two manifolds with boundaries can be glued together along a boundary. If this is done the right way, the result is also a manifold. Similarly, two boundaries of a single manifold can be glued together.

Formally, the gluing is defined by a bijection between the two boundaries. Two points are identified when they are mapped onto each other. For a topological manifold this bijection should be a homeomorphism, otherwise the result will not be a topological manifold. Similarly for a differentiable manifold it has to be a diffeomorphism. For other manifolds other structures should be preserved.

A finite cylinder may be constructed as a manifold by starting with a strip [0, 1] × [0, 1] and gluing a pair of opposite edges on the boundary by a suitable diffeomorphism. A projective plane may be obtained by gluing a sphere with a hole in it to a Möbius strip along their respective circular boundaries.

Cartesian Products

The Cartesian product of manifolds is also a manifold.

The dimension of the product manifold is the sum of the dimensions of its factors. Its topology is the product topology, and a Cartesian product of charts is a chart for the product manifold. Thus, an atlas for the product manifold can be constructed using atlases for its factors. If these atlases define a differential structure on the factors, the corresponding atlas defines a differential structure on the product manifold. The same is true for any other structure defined on the factors. If one of the factors has a boundary, the product manifold also has a boundary. Cartesian products may be used to construct tori and finite cylinders, for example, as $S^1 \times S^1$ and $S^1 \times [0, 1]$, respectively.

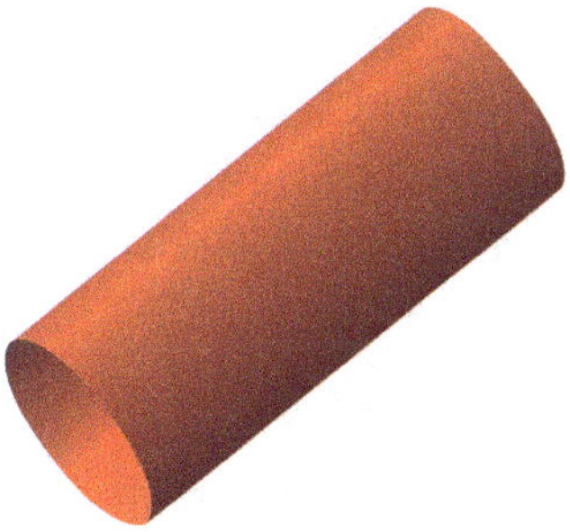

A finite cylinder is a manifold with boundary.

History

The study of manifolds combines many important areas of mathematics: it generalizes concepts such as curves and surfaces as well as ideas from linear algebra and topology.

Early Development

Before the modern concept of a manifold there were several important results.

Non-Euclidean geometry considers spaces where Euclid's parallel postulate fails. Saccheri first studied such geometries in 1733 but sought only to disprove them. Gauss, Bolyai and Lobachevsky independently discovered them 100 years later. Their research uncovered two types of spaces whose geometric structures differ from that of classical Euclidean space; these gave rise to hyperbolic geometry and elliptic geometry. In the modern theory of manifolds, these notions correspond to Riemannian manifolds with constant negative and positive curvature, respectively.

Carl Friedrich Gauss may have been the first to consider abstract spaces as mathematical objects in their own right. His theorema egregium gives a method for computing the curvature of a surface without considering the ambient space in which the surface lies. Such a surface would, in modern terminology, be called a manifold; and in modern terms, the theorem proved that the curvature of the surface is an intrinsic property. Manifold theory has come to focus exclusively on these intrinsic properties (or invariants), while largely ignoring the extrinsic properties of the ambient space.

Another, more topological example of an intrinsic property of a manifold is its Euler characteristic. Leonhard Euler showed that for a convex polytope in the three-dimensional Euclidean space with V vertices (or corners), E edges, and F faces,

$$V - E + F = 2.$$

The same formula will hold if we project the vertices and edges of the polytope onto a sphere, creating a topological map with V vertices, E edges, and F faces, and in fact, will remain true for any spherical map, even if it does not arise from any convex polytope. Thus 2 is a topological invariant of the sphere, called its Euler characteristic. On the other hand, a torus can be sliced open by its 'parallel' and 'meridian' circles, creating a map with $V = 1$ vertex, $E = 2$ edges, and $F = 1$ face. Thus the Euler characteristic of the torus is $1 - 2 + 1 = 0$. The Euler characteristic of other surfaces is a useful topological invariant, which can be extended to higher dimensions using Betti numbers. In the mid nineteenth century, the Gauss–Bonnet theorem linked the Euler characteristic to the Gaussian curvature.

Synthesis

Investigations of Niels Henrik Abel and Carl Gustav Jacobi on inversion of elliptic integrals in the first half of 19th century led them to consider special types of complex manifolds, now known as Jacobians. Bernhard Riemann further contributed to their theory, clarifying the geometric meaning of the process of analytic continuation of functions of complex variables.

Another important source of manifolds in 19th century mathematics was analytical mechanics, as developed by Siméon Poisson, Jacobi, and William Rowan Hamilton. The possible states of a mechanical system are thought to be points of an abstract space, phase space in Lagrangian and Hamiltonian formalisms of classical mechanics. This space is, in fact, a high-dimensional manifold, whose dimension corresponds to the degrees of freedom of the system and where the points are specified by their generalized coordinates. For an unconstrained movement of free particles the manifold is equivalent to the Euclidean space, but various conservation laws constrain it to more complicated formations, e.g. Liouville tori. The theory of a rotating solid body, developed in the 18th century by Leonhard Euler and Joseph-Louis Lagrange, gives another example where the manifold is nontrivial. Geometrical and topological aspects of classical mechanics were emphasized by Henri Poincaré, one of the founders of topology.

Riemann was the first one to do extensive work generalizing the idea of a surface to higher dimensions. The name *manifold* comes from Riemann's original German term, *Mannigfaltigkeit*, which William Kingdon Clifford translated as "manifoldness". In his Göttingen inaugural lecture, Riemann described the set of all possible values of a variable with certain constraints as a *Mannigfaltigkeit*, because the variable can have *many* values. He distinguishes between *stetige Mannigfaltigkeit* and *diskrete Mannigfaltigkeit* (*continuous manifoldness* and *discontinuous manifoldness*), depending on whether the value changes continuously or not. As continuous examples, Riemann refers to not only colors and the locations of objects in space, but also the possible shapes of a spatial figure. Using induction, Riemann constructs an *n-fach ausgedehnte Mannigfaltigkeit* (*n times extended manifoldness* or *n-dimensional manifoldness*) as a continuous stack of (n−1) dimensional manifoldnesses. Riemann's intuitive notion of a *Mannigfaltigkeit* evolved into what is today formalized as a manifold. Riemannian manifolds and Riemann surfaces are named after Riemann.

Poincaré's Definition

In his very influential paper, Analysis Situs, Henri Poincaré gave a definition of a (differentiable) manifold (*variété*) which served as a precursor to the modern concept of a manifold.

In the first section of Analysis Situs, Poincaré defines a manifold as the level set of a continuously differentiable function between Euclidean spaces that satisfies the nondegeneracy hypothesis of the implicit function theorem. In the third section, he begins by remarking that the graph of a continuously differentiable function is a manifold in the latter sense. He then proposes a new, more general, definition of manifold based on a 'chain of manifolds' (*une chaîne des variétés*).

Poincaré's notion of a *chain of manifolds* is a precursor to the modern notion of atlas. In particular, he considers two manifolds defined respectively as graphs of functions $\theta(y)$ and $\theta'(y')$. If these manifolds overlap (*a une partie commune*), then he requires that the coordinates y depend continuously differentiably on the coordinates y' and vice versa ('*...les y sont fonctions analytiques des y' et inversement*'). In this way he introduces a precursor to the notion of a chart and of a transition map. Note that it is implicit in Analysis Situs that a manifold obtained as a 'chain' is a subset of Euclidean space.

For example, the unit circle in the plane can be thought of as the graph of the function $y=\sqrt{1-x^2}$ or else the function $y=-\sqrt{1-x^2}$ in a neighborhood of every point except the points (1, 0) and (−1, 0); and in a neighborhood of those points, it can be thought of as the graph of, respectively, $x=\sqrt{1-y^2}$ and $x=-\sqrt{1-y^2}$. The reason the circle can be represented by a graph in the neighborhood of every point is because the left hand side of its defining equation $x^2+y^2-1=0$ has nonzero gradient at every point of the circle. By the implicit function theorem, every submanifold of Euclidean space is locally the graph of a function.

Hermann Weyl gave an intrinsic definition for differentiable manifolds in his lecture course on Riemann surfaces in 1911–1912, opening the road to the general concept of a topological space that followed shortly. During the 1930s Hassler Whitney and others clarified the foundational aspects of the subject, and thus intuitions dating back to the latter half of the 19th century became precise, and developed through differential geometry and Lie group theory. Notably, the Whitney embedding theorem showed that the intrinsic definition in terms of charts was equivalent to Poincaré's definition in terms of subsets of Euclidean space.

Topology of Manifolds: Highlights

Two-dimensional manifolds, also known as a 2D *surfaces* embedded in our common 3D space, were considered by Riemann under the guise of Riemann surfaces, and rigorously classified in the beginning of the 20th century by Poul Heegaard and Max Dehn. Henri Poincaré pioneered the study of three-dimensional manifolds and raised a fundamental question about them, today known as the Poincaré conjecture. After nearly a century of effort by many mathematicians, starting with Poincaré himself, a consensus among experts (as of 2006) is that Grigori Perelman has proved the Poincaré conjecture. William Thurston's geometrization program, formulated in the 1970s, provided a far-reaching extension of the Poincaré conjecture to the general three-dimensional manifolds. Four-dimensional manifolds were brought to the forefront of mathematical research in the 1980s by Michael Freedman and in a different setting, by Simon Donaldson, who was motivated by the then recent progress in theoretical physics (Yang–Mills theory), where they serve as a substitute for ordinary 'flat' spacetime. Andrey Markov Jr. showed in 1960 that no algorithm exists for classifying four-dimensional manifolds. Important work on higher-dimensional manifolds, including analogues of the Poincaré conjecture, had been done earlier by René Thom, John Milnor, Stephen Smale and Sergei Novikov. One of the most pervasive and flexible techniques underlying much work on the topology of manifolds is Morse theory.

Additional Structure

Topological Manifolds

The simplest kind of manifold to define is the topological manifold, which looks locally like some "ordinary" Euclidean space R^n. Formally, a topological manifold is a topological space locally homeomorphic to a Euclidean space. This means that every point has a neighbourhood for which there exists a homeomorphism (a bijective continuous function whose inverse is also continuous) mapping that neighbourhood to R^n. These homeomorphisms are the charts of the manifold.

It is to be noted that a *topological* manifold looks locally like a Euclidean space in a rather weak manner: while for each individual chart it is possible to distinguish differentiable functions or measure distances and angles, merely by virtue of being a topological manifold a space does not have any *particular* and *consistent* choice of such concepts. In order to discuss such properties for a manifold, one needs to specify further structure and consider differentiable manifolds and Riemannian manifolds discussed below. In particular, the same underlying topological manifold can have several mutually incompatible classes of differentiable functions and an infinite number of ways to specify distances and angles.

Usually additional technical assumptions on the topological space are made to exclude pathological cases. It is customary to require that the space be Hausdorff and second countable.

The *dimension* of the manifold at a certain point is the dimension of the Euclidean space that the charts at that point map to (number n in the definition). All points in a connected manifold have the same dimension. Some authors require that all charts of a topological manifold map to Euclidean spaces of same dimension. In that case every topological manifold has a topological invariant, its dimension. Other authors allow disjoint unions of topological manifolds with differing dimensions to be called manifolds.

Differentiable Manifolds

For most applications a special kind of topological manifold, namely a differentiable manifold, is used. If the local charts on a manifold are compatible in a certain sense, one can define directions, tangent spaces, and differentiable functions on that manifold. In particular it is possible to use calculus on a differentiable manifold. Each point of an n-dimensional differentiable manifold has a tangent space. This is an n-dimensional Euclidean space consisting of the tangent vectors of the curves through the point.

Two important classes of differentiable manifolds are smooth and analytic manifolds. For smooth manifolds the transition maps are smooth, that is infinitely differentiable. Analytic manifolds are smooth manifolds with the additional condition that the transition maps are analytic (they can be expressed as power series). The sphere can be given analytic structure, as can most familiar curves and surfaces.

There are also topological manifolds, i.e., locally Euclidean spaces, which possess no differentiable structures at all.

A rectifiable set generalizes the idea of a piecewise smooth or rectifiable curve to higher dimensions; however, rectifiable sets are not in general manifolds.

Riemannian Manifolds

To measure distances and angles on manifolds, the manifold must be Riemannian. A 'Riemannian manifold' is a differentiable manifold in which each tangent space is equipped with an inner product $\langle\cdot,\cdot\rangle$ in a manner which varies smoothly from point to point. Given two tangent vectors u and v, the inner product $\langle u,v\rangle$ gives a real number.

The dot (or scalar) product is a typical example of an inner product. This allows one to define various notions such as length, angles, areas (or volumes), curvature and divergence of vector fields.

All differentiable manifolds (of constant dimension) can be given the structure of a Riemannian manifold. The Euclidean space itself carries a natural structure of Riemannian manifold (the tangent spaces are naturally identified with the Euclidean space itself and carry the standard scalar product of the space). Many familiar curves and surfaces, including for example all n-spheres, are specified as subspaces of a Euclidean space and inherit a metric from their embedding in it.

Finsler Manifolds

A Finsler manifold allows the definition of distance but does not require the concept of angle; it is an analytic manifold in which each tangent space is equipped with a norm, $||\cdot||$, in a manner which varies smoothly from point to point. This norm can be extended to a metric, defining the length of a curve; but it cannot in general be used to define an inner product.

Any Riemannian manifold is a Finsler manifold.

Lie Groups

Lie groups, named after Sophus Lie, are differentiable manifolds that carry also the structure of a group which is such that the group operations are defined by smooth maps.

A Euclidean vector space with the group operation of vector addition is an example of a non-compact Lie group. A simple example of a compact Lie group is the circle: the group operation is simply rotation. This group, known as U(1), can be also characterised as the group of complex numbers of modulus 1 with multiplication as the group operation.

Other examples of Lie groups include special groups of matrices, which are all subgroups of the general linear group, the group of n by n matrices with non-zero determinant. If the matrix entries are real numbers, this will be an n^2-dimensional disconnected manifold. The orthogonal groups, the symmetry groups of the sphere and hyperspheres, are $n(n-1)/2$ dimensional manifolds, where $n-1$ is the dimension of the sphere. Further examples can be found in the table of Lie groups.

Other Types of Manifolds

- A *complex manifold* is a manifold whose charts take values in $\mathbb{C}^n$ and whose transition functions are holomorphic on the overlaps. These manifolds are the basic objects of study in complex geometry. A one-complex-dimensional manifold is called a Riemann surface. Note that an n-dimensional complex manifold has dimension $2n$ as a real differentiable manifold.
- A *CR manifold* is a manifold modeled on boundaries of domains in $\mathbb{C}^n$.

- 'Infinite dimensional manifolds': to allow for infinite dimensions, one may consider Banach manifolds which are locally homeomorphic to Banach spaces. Similarly, Fréchet manifolds are locally homeomorphic to Fréchet spaces.
- A *symplectic manifold* is a kind of manifold which is used to represent the phase spaces in classical mechanics. They are endowed with a 2-form that defines the Poisson bracket. A closely related type of manifold is a contact manifold.
- A *combinatorial manifold* is a kind of manifold which is discretization of a manifold. It usually means a piecewise linear manifold made by simplicial complexes.
- A *digital manifold* is a special kind of combinatorial manifold which is defined in digital space.

Classification and Invariants

Different notions of manifolds have different notions of classification and invariant; in this section we focus on smooth closed manifolds.

The classification of smooth closed manifolds is well-understood *in principle*, except in dimension 4: in low dimensions (2 and 3) it is geometric, via the uniformization theorem and the solution of the Poincaré conjecture, and in high dimension (5 and above) it is algebraic, via surgery theory. This is a classification in principle: the general question of whether two smooth manifolds are diffeomorphic is not computable in general. Further, specific computations remain difficult, and there are many open questions.

Orientable surfaces can be visualized, and their diffeomorphism classes enumerated, by genus. Given two orientable surfaces, one can determine if they are diffeomorphic by computing their respective genera and comparing: they are diffeomorphic if and only if the genera are equal, so the genus forms a complete set of invariants.

This is much harder in higher dimensions: higher-dimensional manifolds cannot be directly visualized (though visual intuition is useful in understanding them), nor can their diffeomorphism classes be enumerated, nor can one in general determine if two different descriptions of a higher-dimensional manifold refer to the same object.

However, one can determine if two manifolds are *different* if there is some intrinsic characteristic that differentiates them. Such criteria are commonly referred to as invariants, because, while they may be defined in terms of some presentation (such as the genus in terms of a triangulation), they are the same relative to all possible descriptions of a particular manifold: they are *invariant* under different descriptions.

Naively, one could hope to develop an arsenal of invariant criteria that would definitively classify all manifolds up to isomorphism. Unfortunately, it is known that for manifolds of dimension 4 and higher, no program exists that can decide whether two manifolds are diffeomorphic.

Smooth manifolds have a rich set of invariants, coming from point-set topology, classic algebraic topology, and geometric topology. The most familiar invariants, which are visible for surfaces, are orientability (a normal invariant, also detected by homology) and genus (a homological invariant).

Smooth closed manifolds have no local invariants (other than dimension), though geometric manifolds have local invariants, notably the curvature of a Riemannian manifold and the torsion of a manifold equipped with an affine connection. This distinction between local invariants and no local invariants is a common way to distinguish between geometry and topology. All invariants of a smooth closed manifold are thus global.

Algebraic topology is a source of a number of important global invariant properties. Some key criteria include the *simply connected* property and orientability. Indeed, several branches of mathematics, such as homology and homotopy theory, and the theory of characteristic classes were founded in order to study invariant properties of manifolds.

Surfaces

Orientability

In dimensions two and higher, a simple but important invariant criterion is the question of whether a manifold admits a meaningful orientation. Consider a topological manifold with charts mapping to R^n. Given an ordered basis for R^n, a chart causes its piece of the manifold to itself acquire a sense of ordering, which in 3-dimensions can be viewed as either right-handed or left-handed. Overlapping charts are not required to agree in their sense of ordering, which gives manifolds an important freedom. For some manifolds, like the sphere, charts can be chosen so that overlapping regions agree on their "handedness"; these are *orientable* manifolds. For others, this is impossible. The latter possibility is easy to overlook, because any closed surface embedded (without self-intersection) in three-dimensional space is orientable.

Some illustrative examples of non-orientable manifolds include: (1) the Möbius strip, which is a manifold with boundary, (2) the Klein bottle, which must intersect itself in its 3-space representation, and (3) the real projective plane, which arises naturally in geometry.

Möbius Strip

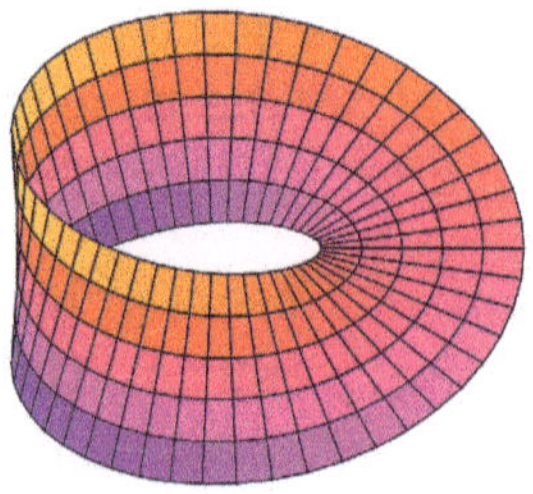

Möbius strip

Begin with an infinite circular cylinder standing vertically, a manifold without boundary. Slice across it high and low to produce two circular boundaries, and the cylindrical strip between them. This is an orientable manifold with boundary, upon which "surgery" will be performed. Slice the strip open, so that it could unroll to become a rectangle, but keep a grasp on the cut ends. Twist one end 180°, making the inner surface face out, and glue the ends back together seamlessly. This results in a strip with a permanent half-twist: the Möbius strip. Its boundary is no longer a pair of circles, but (topologically) a single circle; and what was once its "inside" has merged with its "outside", so that it now has only a *single* side.

Klein Bottle

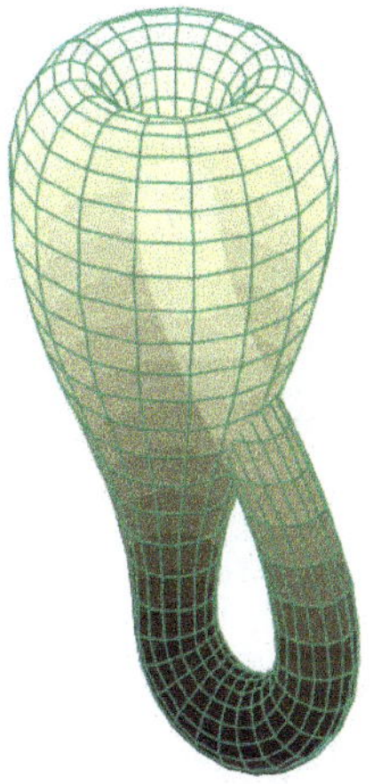

The Klein bottle immersed in three-dimensional space

Take two Möbius strips; each has a single loop as a boundary. Straighten out those loops into circles, and let the strips distort into cross-caps. Gluing the circles together will produce a new, closed manifold without boundary, the Klein bottle. Closing the surface does nothing to improve the lack of orientability, it merely removes the boundary. Thus, the Klein bottle is a closed surface with no distinction between inside and outside. Note that in three-dimensional space, a Klein bottle's surface must pass through itself. Building a Klein bottle which is not self-intersecting requires four or more dimensions of space.

Real Projective Plane

Begin with a sphere centered on the origin. Every line through the origin pierces the sphere in two opposite points called *antipodes*. Although there is no way to do so physically, it is possible (by considering a quotient space) to mathematically merge each antipode pair into a single point. The closed surface so produced is the real projective plane, yet another non-orientable surface. It has a number of equivalent descriptions and constructions, but this route explains its name: all the points on any given line through the origin project to the same "point" on this "plane".

Genus and the Euler Characteristic

For two dimensional manifolds a key invariant property is the genus, or the "number of handles" present in a surface. A torus is a sphere with one handle, a double torus is a sphere with two handles, and so on. Indeed, it is possible to fully characterize compact, two-dimensional manifolds on the basis of genus and orientability. In higher-dimensional manifolds genus is replaced by the notion of Euler characteristic, and more generally Betti numbers and homology and cohomology.

Maps of Manifolds

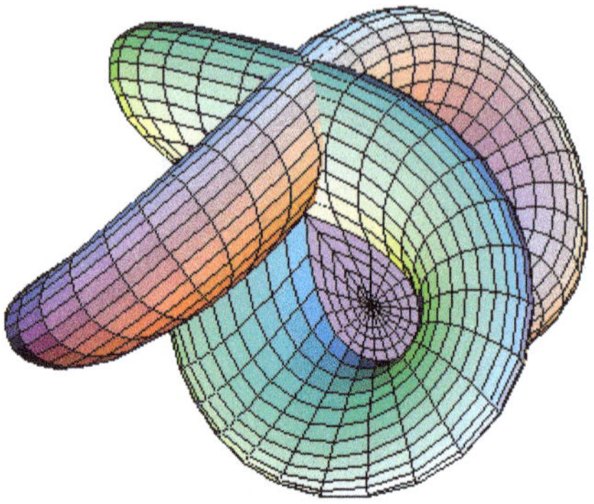

A Morin surface, an immersion used in sphere eversion

Just as there are various types of manifolds, there are various types of maps of manifolds. In addition to continuous functions and smooth functions generally, there are maps with special properties. In geometric topology a basic type are embeddings, of which knot theory is a central example, and generalizations such as immersions, submersions, covering spaces, and ramified covering spaces. Basic results include the Whitney embedding theorem and Whitney immersion theorem.

In Riemannian geometry, one may ask for maps to preserve the Riemannian metric, leading to notions of isometric embeddings, isometric immersions, and Riemannian submersions; a basic result is the Nash embedding theorem.

Scalar-valued Functions

A basic example of maps between manifolds are scalar-valued functions on a manifold,

$$f : M \to \mathrm{R} \text{ or } f : M \to \mathrm{C},$$

sometimes called regular functions or functionals, by analogy with algebraic geometry or linear algebra. These are of interest both in their own right, and to study the underlying manifold.

In geometric topology, most commonly studied are Morse functions, which yield handlebody decompositions, while in mathematical analysis, one often studies solution to partial differential equations, an important example of which is harmonic analysis, where one studies harmonic functions: the kernel of the Laplace operator. This leads to such functions as the spherical harmonics, and to heat kernel methods of studying manifolds, such as hearing the shape of a drum and some proofs of the Atiyah–Singer index theorem.

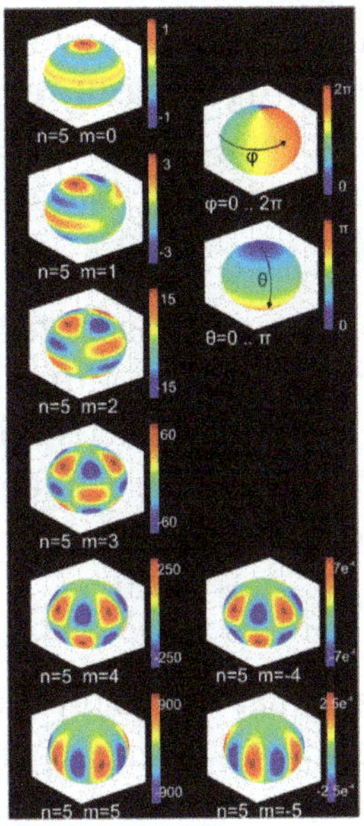

3D color plot of the spherical harmonics of degree $n = 5$

Generalizations of Manifolds

Infinite dimensional manifolds

The definition of a manifold can be generalized by dropping the requirement of finite dimensionality. Thus an infinite dimensional manifold is a topological space locally homeomorphic to a topological vector space over the reals. This omits the point-set axioms, allowing higher cardinalities and non-Hausdorff manifolds; and it omits finite dimension, allowing structures such as Hilbert manifolds to be modeled on Hilbert spaces, Banach manifolds to be modeled on Banach spaces, and Fréchet manifolds to be modeled on Fréchet spaces. Usually one relaxes one or the other condition: manifolds with the point-set axioms are studied in general topology, while infinite-dimensional manifolds are studied in functional analysis.

Orbifolds

An orbifold is a generalization of manifold allowing for certain kinds of "singularities" in the topology. Roughly speaking, it is a space which locally looks like the quotients of some simple space (*e.g.* Euclidean space) by the actions of various finite groups. The singularities correspond to fixed points of the group actions, and the actions must be compatible in a certain sense.

Algebraic varieties and schemes

Non-singular algebraic varieties over the real or complex numbers are manifolds. One generalizes this first by allowing singularities, secondly by allowing different fields, and thirdly by emulating the patching construction of manifolds: just as a manifold is glued together from open subsets of Euclidean space, an algebraic variety is glued together from affine algebraic varieties, which are zero sets of polynomials over algebraically closed fields. Schemes are likewise glued together from affine schemes, which are a generalization of algebraic varieties. Both are related to manifolds, but are constructed algebraically using sheaves instead of atlases.

Because of singular points, a variety is in general not a manifold, though linguistically the French *variété*, German *Mannigfaltigkeit* and English *manifold* are largely synonymous. In French an algebraic variety is called *une variété algébrique* (an *algebraic variety*), while a smooth manifold is called *une variété différentielle* (a *differential variety*).

Stratified space

A "stratified space" is a space that can be divided into pieces ("strata"), with each stratum a manifold, with the strata fitting together in prescribed ways (formally, a filtration by closed subsets). There are various technical definitions, notably a Whitney stratified space for smooth manifolds and a topologically stratified space for topological manifolds. Basic examples include manifold with boundary (top dimensional manifold and codimension 1 boundary) and manifold with corners (top dimensional manifold, codimension 1 boundary, codimension 2 corners). Whitney stratified spaces are a broad class of spaces, including algebraic varieties, analytic varieties, semialgebraic sets, and subanalytic sets.

CW-complexes

A CW complex is a topological space formed by gluing disks of different dimensionality together. In general the resulting space is singular, and hence not a manifold. However, they are of central interest in algebraic topology, especially in homotopy theory, as they are easy to compute with and singularities are not a concern.

Homology manifolds

A homology manifold is a space that behaves like a manifold from the point of view of homology theory. These are not all manifolds, but (in high dimension) can be analyzed by surgery theory similarly to manifolds, and failure to be a manifold is a local obstruction, as in surgery theory.

Differential spaces

Let M be a nonempty set. Suppose that some family of real functions on M was chosen. Denote it by $C \subseteq \mathbb{R}^M$. It is an algebra with respect to the pointwise addition and multiplication. Let M be equipped with the topology induced by C. Suppose also that the following conditions hold. First: for every $H \in C^\infty(\mathbb{R}^i)$, where $i \in \mathbb{N}$, and arbitrary $f_1, \ldots, f_n \in C$, the composition $H \circ (f_1, \ldots, f_n) \in C$. Second: every function, which in every point of M locally coincides with some function from C, also belongs to C. A pair (M, C) for which the above conditions hold, is called a Sikorski differential space.

Space (Mathematics)

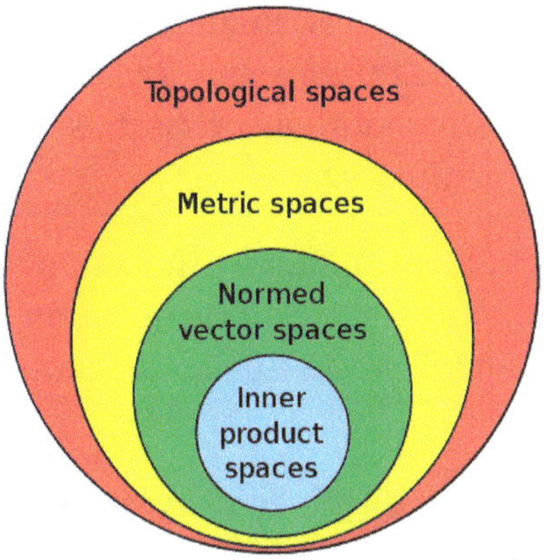

A hierarchy of mathematical spaces: The inner product induces a norm. The norm induces a metric. The metric induces a topology.

In mathematics, a space is a set (sometimes called a universe) with some added structure.

Mathematical spaces often form a hierarchy, i.e., one space may inherit all the characteristics of a parent space. For instance, all inner product spaces are also normed vector spaces, because the inner product *induces* a norm on the inner product space such that:

$$\|x\| = \sqrt{\langle x, x \rangle},$$

where the norm is indicated by enclosing in double vertical lines, and the inner product is indicated enclosing in by angle brackets.

Modern mathematics treats "space" quite differently compared to classical mathematics.

History

Before the Golden Age of Geometry

In the ancient mathematics, "space" was a geometric abstraction of the three-dimensional space observed in the everyday life. The axiomatic method had been the main research tool since Euclid (about 300 BC). The method of coordinates (analytic geometry) was adopted by René Descartes in 1637. At that time, geometric theorems were treated as an absolute objective truth knowable through intuition and reason, similar to objects of natural science; and axioms were treated as obvious implications of definitions.

Two equivalence relations between geometric figures were used: congruence and similarity. Translations, rotations and reflections transform a figure into congruent figures; homotheties – into similar figures. For example, all circles are mutually similar, but ellipses are not similar to circles. A third equivalence relation, introduced by Gaspard Monge in 1795, occurs in projective geometry: not only ellipses, but also parabolas and hyperbolas, turn into circles under appropriate projective transformations; they all are projectively equivalent figures.

The relation between the two geometries, Euclidean and projective, shows that mathematical objects are not given to us *with their structure*. Rather, each mathematical theory describes its objects by *some* of their properties, precisely those that are put as axioms at the foundations of the theory.

Distances and angles are never mentioned in the axioms of the projective geometry and therefore cannot appear in its theorems. The question "what is the sum of the three angles of a triangle" is meaningful in the Euclidean geometry but meaningless in the projective geometry.

A different situation appeared in the 19th century: in some geometries the sum of the three angles of a triangle is well-defined but different from the classical value (180 degrees). The non-Euclidean hyperbolic geometry, introduced by Nikolai Lobachevsky in 1829 and János Bolyai in 1832 (and Carl Gauss in 1816, unpublished) stated that the sum depends on the triangle and is always less than 180 degrees. Eugenio Beltrami in 1868 and Felix Klein in 1871 obtained Euclidean "models" of the non-Euclidean hyperbolic geometry, and thereby completely justified this theory.

This discovery forced the abandonment of the pretensions to the absolute truth of Euclidean geometry. It showed that axioms are not "obvious", nor "implications of definitions". Rather, they are hypotheses. To what extent do they correspond to an experimental reality? This important physical problem no longer has anything to do with mathematics. Even if a "geometry" does not correspond to an experimental reality, its theorems remain no less "mathematical truths".

A Euclidean model of a non-Euclidean geometry is a clever choice of some objects existing in Euclidean space and some relations between these objects that satisfy all axioms (therefore, all theorems) of the non-Euclidean geometry. These Euclidean objects and relations "play" the non-Euclidean geometry like contemporary actors playing an ancient performance. Relations between the actors only mimic relations between the characters in the play. Likewise, the chosen relations between the chosen objects of the Euclidean model only mimic the non-Euclidean relations. It shows that relations between objects are essential in mathematics, while the nature of the objects is not.

The Golden Age and Afterwards: Dramatic Change

According to Nicolas Bourbaki, the period between 1795 (*Geometrie descriptive* of Monge) and 1872 (the "Erlangen programme" of Klein) can be called the golden age of geometry. Analytic geometry made a great progress and succeeded in replacing theorems of classical geometry with computations via invariants of transformation groups. Since that time new theorems of classical geometry are of more interest to amateurs rather than to professional mathematicians.

However, it does not mean that the heritage of the classical geometry was lost. According to Bourbaki, "passed over in its role as an autonomous and living science, classical geometry is thus transfigured into a universal language of contemporary mathematics".

According to the famous inaugural lecture given by Bernhard Riemann in 1854, every mathematical object parametrized by n real numbers may be treated as a point of the n-dimensional space of all such objects. Contemporary mathematicians follow this idea routinely and find it extremely suggestive to use the terminology of classical geometry nearly everywhere.

In order to fully appreciate the generality of this approach one should note that mathematics is "a pure theory of forms, which has as its purpose, not the combination of quantities, or of their images, the numbers, but objects of thought" (Hermann Hankel, 1867). This is a controversial characterization of the purpose of mathematics, which is not necessarily committed to the existence of "objects of thought."

Functions are important mathematical objects. Usually they form infinite-dimensional function spaces, as noted already by Riemann and elaborated in the 20th century by functional analysis.

An object parametrized by n complex numbers may be treated as a point of a complex n-dimensional space. However, the same object is also parametrized by 2 n real numbers (if c is a complex number, then $c = a + b$ i, where a and b are real), thus, a point of a real 2n-dimensional space. The complex dimension differs from the real dimension. This is only the tip of the iceberg. The "algebraic" concept of dimension applies to vector spaces. For topological spaces there are several dimension concepts including inductive dimension and Hausdorff dimension, which can be non-integer (especially for fractals). Some kinds of spaces (for instance, measure spaces) admit no concept of dimension at all.

The original space investigated by Euclid is now called three-dimensional Euclidean space. Its axiomatization, started by Euclid 23 centuries ago, was reformed with Hilbert's axioms, Tarski's axioms and Birkhoff's axioms. These axiom systems describe the space via primitive notions (such as "point", "between", "congruent") constrained by a number of axioms. Such a definition "from scratch" is now not often used, since it does not reveal the relation of this space to other spaces. The modern approach defines the three-dimensional Euclidean space more algebraically, via vector spaces and quadratic forms, namely, as an affine space whose difference space is a three-dimensional inner product space.

Also a three-dimensional projective space is now defined non-classically, as the space of all one-dimensional subspaces (that is, straight lines through the origin) of a four-dimensional vector space.

A space consists now of selected mathematical objects (for instance, functions on another space, or subspaces of another space, or just elements of a set) treated as points, and selected relationships between these points. It shows that spaces are just mathematical structures of convenience. One may expect that the structures called "spaces" are more geometric than others, but this is not always true. For example, a differentiable manifold (called also smooth manifold) is much more geometric than a measurable space, but no one calls it "differentiable space" (nor "smooth space").

Taxonomy of Spaces

Three Taxonomic Ranks

Spaces are classified on three levels. Given that each mathematical theory describes its objects by *some* of their properties, the first question to ask is: which properties?

For example, the upper-level classification distinguishes between Euclidean and projective spaces, since the distance between two points is defined in Euclidean spaces but undefined in projective spaces. These are spaces of different types.

Another example. The question "what is the sum of the three angles of a triangle" makes sense in a Euclidean space but not in a projective space; these are spaces of different types. In a non-Euclidean space the question makes sense but is answered differently, which is not an upper-level distinction.

Also, the distinction between a Euclidean plane and a Euclidean 3-dimensional space is not an upper-level distinction; the question "what is the dimension" makes sense in both cases.

In terms of Bourbaki the upper-level classification is related to "typical characterization" (or "typification"). However, it is not the same (since two equivalent structures may differ in typification).

On the second level of classification one takes into account answers to especially important questions (among the questions that make sense according to the first level). For example, this level distinguishes between Euclidean and non-Euclidean spaces; between finite-dimensional and infinite-dimensional spaces; between compact and non-compact spaces, etc.

In terms of Bourbaki the second-level classification is the classification by "species". Unlike biological taxonomy, a space may belong to several species.

On the third level of classification, roughly speaking, one takes into account answers to *all possible* questions (that make sense according to the first level). For example, this level distinguishes between spaces of different dimension, but does not distinguish between a plane of a three-dimensional Euclidean space, treated as a two-dimensional Euclidean space, and the set of all pairs of real numbers, also treated as a two-dimensional Euclidean space. Likewise it does not distinguish between different Euclidean models of the same non-Euclidean space.

More formally, the third level classifies spaces up to isomorphism. An isomorphism between two spaces is defined as a one-to-one correspondence between the points of the first space and the points of the second space, that preserves all relations between the points, stipulated by the given "typification". Mutually isomorphic spaces are thought of as copies of a single space. If one of them belongs to a given species then they all do.

The notion of isomorphism sheds light on the upper-level classification. Given a one-to-one correspondence between two spaces of the same type, one may ask whether it is an isomorphism or not. This question makes no sense for two spaces of different type.

Isomorphisms to itself are called automorphisms. Automorphisms of a Euclidean space are motions and reflections. Euclidean space is homogeneous in the sense that every point can be transformed into every other point by some automorphism.

Two Relations between Spaces, and a Property of Spaces

Topological notions (continuity, convergence, open sets, closed sets etc.) are defined naturally in every Euclidean space. In other words, every Euclidean space is also a topological space. Every isomorphism between two Euclidean spaces is also an isomorphism between the corresponding topological spaces (called "homeomorphism"), but the converse is wrong: a homeomorphism may distort distances. In terms of Bourbaki, "topological space" is an underlying structure of the "Euclidean space" structure. Similar ideas occur in category theory: the category of Euclidean spaces is a concrete category over the category of topological spaces; the forgetful (or "stripping") functor maps the former category to the latter category.

A three-dimensional Euclidean space is a special case of a Euclidean space. In terms of Bourbaki, the species of three-dimensional Euclidean space is richer than the species of Euclidean space. Likewise, the species of compact topological space is richer than the species of topological space.

Euclidean axioms leave no freedom, they determine uniquely all geometric properties of the space. More exactly: all three-dimensional Euclidean spaces are mutually isomorphic. In this sense we have "the" three-dimensional Euclidean space. In terms of Bourbaki, the corresponding theory is univalent. In contrast, topological spaces are generally non-isomorphic, their theory is multivalent. A similar idea occurs in mathematical logic: a theory is called categorical if all its models of the same cardinality are mutually isomorphic. According to Bourbaki, the study of multivalent theories is the most striking feature which distinguishes modern mathematics from classical mathematics.

Types of Spaces

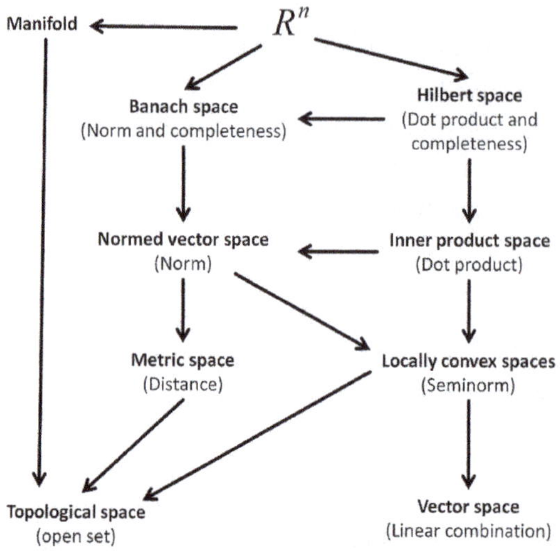

Overview of types of abstract spaces. An arrow from space A to space B implies that space A is also a kind of space B. That means, for instance, that a normed vector space is also a metric space.

Linear and Topological Spaces

Two basic spaces are linear spaces (also called vector spaces) and topological spaces.

Linear spaces are of algebraic nature; there are real linear spaces (over the field of real numbers), complex linear spaces (over the field of complex numbers), and more generally, linear spaces over any field. Every complex linear space is also a real linear space (the latter *underlies* the former), since each real number is also a complex number. Linear operations, given in a linear space by definition, lead to such notions as straight lines (and planes, and other linear subspaces); parallel lines; ellipses (and ellipsoids). However, orthogonal (perpendicular) lines cannot be defined, and circles cannot be singled out among ellipses. The dimension of a linear space is defined as the maximal number of linearly independent vectors or, equivalently, as the minimal number of vectors that span the space; it may be finite or infinite. Two linear spaces over the same field are isomorphic if and only if they are of the same dimension.

Topological spaces are of analytic nature. Open sets, given in a topological space by definition, lead to such notions as continuous functions, paths, maps; convergent sequences, limits; interior, boundary, exterior. However, uniform continuity, bounded sets, Cauchy sequences, differentiable functions (paths, maps) remain undefined. Isomorphisms between topological spaces are traditionally called homeomorphisms; these are one-to-one correspondences continuous in both directions. The open interval $(0,1)$ is homeomorphic to the whole real line $(-\infty,\infty)$ but not homeomorphic to the closed interval $[0,1]$, nor to a circle. The surface of a cube is homeomorphic to a sphere (the surface of a ball) but not homeomorphic to a torus. Euclidean spaces of different dimensions are not homeomorphic, which seems evident, but is not easy to prove. Dimension of a topological space is difficult to define; "inductive dimension" and "Lebesgue covering dimension" are used. Every subset of a topological space is itself a topological space (in contrast, only *linear* subsets of a linear space are linear spaces). Arbitrary topological spaces, investigated by general topology (called also point-set topology) are too diverse for a complete classification (up to homeomorphism). They are inhomogeneous (in general). Compact topological spaces are an important class of topological spaces ("species" of this "type"). Every continuous function is bounded on such space. The closed interval $[0,1]$ and the extended real line $[-\infty,\infty]$ are compact; the open interval $(0,1)$ and the line $(-\infty,\infty)$ are not. Geometric topology investigates manifolds (another "species" of this "type"); these are topological spaces locally homeomorphic to Euclidean spaces. Low-dimensional manifolds are completely classified (up to homeomorphism).

The two structures discussed above (linear and topological) are both underlying structures of the "linear topological space" structure. That is, a linear topological space is both a linear (real or complex) space and a (homogeneous, in fact) topological space. However, an arbitrary combination of these two structures is generally not a linear topological space; the two structures must conform, namely, the linear operations must be continuous.

Every finite-dimensional (real or complex) linear space is a linear topological space in the sense that it carries one and only one topology that makes it a linear topological space. The two structures, "finite-dimensional (real or complex) linear space" and "finite-dimensional linear topological space", are thus equivalent, that is, mutually underlying. Accordingly, every invertible linear transformation of a finite-dimensional linear topological space is a homeomorphism. In the infinite dimension, however, different topologies conform to a given linear structure, and invertible linear transformations are generally not homeomorphisms.

Affine and Projective Spaces

It is convenient to introduce affine and projective spaces by means of linear spaces, as follows. An n-dimensional linear subspace of an $(n+1)$-dimensional linear space, being itself an n-dimensional linear space, is not homogeneous; it contains a special point, the origin. Shifting it by a vector external to it, one obtains an n-dimensional affine space. It is homogeneous. In the words of John Baez, "an affine space is a vector space that's forgotten its origin". A straight line in the affine space is, by definition, its intersection with a two-dimensional linear subspace (plane through the origin) of the $(n+1)$-dimensional linear space. Every linear space is also an affine space.

Every point of the affine space is its intersection with a one-dimensional linear subspace (line through the origin) of the $(n+1)$-dimensional linear space. However, some one-dimensional subspaces are parallel to the affine space; in some sense, they intersect it at infinity. The set of all one-dimensional linear subspaces of an $(n+1)$-dimensional linear space is, by definition, an n-dimensional projective space. Choosing an n-dimensional affine space as before one observes that the affine space is embedded as a proper subset into the projective space. However, the projective space itself is homogeneous. A straight line in the projective space, by definition, corresponds to a two-dimensional linear subspace of the $(n+1)$-dimensional linear space.

Defined this way, affine and projective spaces are of algebraic nature; they can be real, complex, and more generally, over any field.

Every real (or complex) affine or projective space is also a topological space. An affine space is a non-compact manifold; a projective space is a compact manifold.

Metric and Uniform Spaces

Distances between points are defined in a metric space. Every metric space is also a topological space. Bounded sets and Cauchy sequences are defined in a metric space (but not just in a topological space). Isomorphisms between metric spaces are called isometries. A metric space is called complete if all Cauchy sequences converge. Every incomplete space is isometrically embedded into its completion. Every compact metric space is complete; the real line is non-compact but complete; the open interval $(0,1)$ is incomplete.

A topological space is called metrizable, if it underlies a metric space. All manifolds are metrizable.

Every Euclidean space is also a complete metric space. Moreover, all geometric notions immanent to a Euclidean space can be characterized in terms of its metric. For example, the straight segment connecting two given points A and C consists of all points B such that the distance between A and C is equal to the sum of two distances, between A and B and between B and C.

Uniform spaces do not introduce distances, but still allow one to use uniform continuity, Cauchy sequences, completeness and completion. Every uniform space is also a topological space. Every *linear* topological space (metrizable or not) is also a uniform space. More generally, every commutative topological group is also a uniform space. A non-commutative topological group, however, carries two uniform structures, one left-invariant, the other right-invariant. Linear topological spaces are complete in finite dimension but generally incomplete in infinite dimension.

Normed, Banach, Inner Product, and Hilbert Spaces

Vectors in a Euclidean space are a linear space, but each vector x has also a length, in other words, norm, $\| x \|$. A (real or complex) linear space endowed with a norm is a normed space. Every normed space is both a linear topological space and a metric space. A Banach space is a complete normed space. Many spaces of sequences or functions are infinite-dimensional Banach spaces.

The set of all vectors of norm less than one is called the unit ball of a normed space. It is a convex, centrally symmetric set, generally not an ellipsoid; for example, it may be a polygon (on the plane). The parallelogram law (called also parallelogram identity) $\| x-y \|^2 + \| x+y \|^2 = 2\| x \|^2 + 2\| y \|^2$ generally fails in normed spaces, but holds for vectors in Euclidean spaces, which follows from the fact that the squared Euclidean norm of a vector is its inner product to itself.

An inner product space is a (real or complex) linear space endowed with a bilinear (or sesquilinear) form satisfying some conditions and called inner product. Every inner product space is also a normed space. A normed space underlies an inner product space if and only if it satisfies the parallelogram law, or equivalently, if its unit ball is an ellipsoid. Angles between vectors are defined in inner product spaces. A Hilbert space is defined as a complete inner product space. (Some authors insist that it must be complex, others admit also real Hilbert spaces.) Many spaces of sequences or functions are infinite-dimensional Hilbert spaces. Hilbert spaces are very important for quantum theory.

All n-dimensional real inner product spaces are mutually isomorphic. One may say that the n-dimensional Euclidean space is the n-dimensional real inner product space that's forgotten its origin.

Smooth and Riemannian Manifolds (Spaces)

Smooth manifolds are not called "spaces", but could be. Smooth (differentiable) functions, paths, maps, given in a smooth manifold by definition, lead to tangent spaces. Every smooth manifold is a (topological) manifold. Smooth surfaces in a finite-dimensional linear space (like the surface of an ellipsoid, not a polytope) are smooth manifolds. Every smooth manifold can be embedded into a finite-dimensional linear space. A smooth path in a smooth manifold has (at every point) the tangent vector, belonging to the tangent space (attached to this point). Tangent spaces to an n-dimensional smooth manifold are n-dimensional linear spaces. A smooth function has (at every point) the differential, – a linear functional on the tangent space. Real (or complex) finite-dimensional linear, affine and projective spaces are also smooth manifolds.

A Riemannian manifold, or Riemann space, is a smooth manifold whose tangent spaces are endowed with inner product (satisfying some conditions). Euclidean spaces are also Riemann spaces. Smooth surfaces in Euclidean spaces are Riemann spaces. A hyperbolic non-Euclidean space is also a Riemann space. A curve in a Riemann space has the length. A Riemann space is both a smooth manifold and a metric space; the length of the shortest curve is the distance. The angle between two curves intersecting at a point is the angle between their tangent lines.

Waiving positivity of inner product on tangent spaces one gets pseudo-Riemann (especially, Lorentzian) spaces very important for general relativity.

Measurable, Measure, and Probability Spaces

Waiving distances and angles while retaining volumes (of geometric bodies) one moves toward measure theory. Besides the volume, a measure generalizes area, length, mass (or charge) distribution, and also probability distribution, according to Andrey Kolmogorov's approach to probability theory.

A "geometric body" of classical mathematics is much more regular than just a set of points. The boundary of the body is of zero volume. Thus, the volume of the body is the volume of its interior, and the interior can be exhausted by an infinite sequence of cubes. In contrast, the boundary of an arbitrary set of points can be of non-zero volume (an example: the set of all rational points inside a given cube). Measure theory succeeded in extending the notion of volume (or another measure) to a vast class of sets, so-called measurable sets. Indeed, non-measurable sets almost never occur in applications, but anyway, the theory must restrict itself to measurable sets (and functions).

Measurable sets, given in a measurable space by definition, lead to measurable functions and maps. In order to turn a topological space into a measurable space one endows it with a σ-algebra. The σ-algebra of Borel sets is most popular, but not the only choice (Baire sets, universally measurable sets etc. are used sometimes). Alternatively, a σ-algebra can be generated by a given collection of sets (or functions) irrespective of any topology. Quite often, different topologies lead to the same σ-algebra (for example, the norm topology and the weak topology on a separable Hilbert space). Every subset of a measurable space is itself a measurable space.

Standard measurable spaces (called also standard Borel spaces) are especially useful. Every Borel set (in particular, every closed set and every open set) in a Euclidean space (and more generally, in a complete separable metric space) is a standard measurable space. All uncountable standard measurable spaces are mutually isomorphic.

A measure space is a measurable space endowed with a measure. A Euclidean space with Lebesgue measure is a measure space. Integration theory defines integrability and integrals of measurable functions on a measure space.

Sets of measure 0, called null sets, are negligible. Accordingly, a $\bmod 0$ isomorphism is defined as isomorphism between subsets of full measure (that is, with negligible complement).

A probability space is a measure space such that the measure of the whole space is equal to 1. The product of any family (finite or not) of probability spaces is a probability space. In contrast, for measure spaces in general, only the product of finitely many spaces is defined. Accordingly, there are many infinite-dimensional probability measures (especially, Gaussian measures), but no infinite-dimensional Lebesgue measure.

Standard probability spaces are especially useful. Every probability measure on a standard measurable space leads to a standard probability space. The product of a sequence (finite or not) of standard probability spaces is a standard probability space. All non-atomic standard probability spaces are mutually isomorphic $\bmod 0$ one of them is the interval $(0,1)$ with Lebesgue measure.

These spaces are less geometric. In particular, the idea of dimension, applicable (in one form or another) to all other spaces, does not apply to measurable, measure and probability spaces.

A topological space becomes also a measurable space when endowed with the Borel σ-algebra. However, the topology is not uniquely determined by its Borel σ-algebra; and not every σ-algebra is the Borel σ-algebra of some topology.

Topological Property

In topology and related areas of mathematics a topological property or topological invariant is a property of a topological space which is invariant under homeomorphisms. That is, a property of spaces is a topological property if whenever a space X possesses that property every space homeomorphic to X possesses that property. Informally, a topological property is a property of the space that can be expressed using open sets.

A common problem in topology is to decide whether two topological spaces are homeomorphic or not. To prove that two spaces are *not* homeomorphic, it is sufficient to find a topological property which is not shared by them.

Common Topological Properties

Cardinal Functions

- The cardinality $|X|$ of the space X.
- The cardinality $\tau(X)$ of the topology of the space X.
- *Weight* $w(X)$, the least cardinality of a basis of the topology of the space X.
- *Density* $d(X)$, the least cardinality of a subset of X whose closure is X.

Separation

- T_0 or Kolmogorov. A space is Kolmogorov if for every pair of distinct points x

and y in the space, there is at least either an open set containing x but not y, or an open set containing y but not x.

- T_1 or Fréchet. A space is Fréchet if for every pair of distinct points x and y in the space, there is an open set containing x but not y. (Compare with T_0; here, we are allowed to specify which point will be contained in the open set.) Equivalently, a space is T_1 if all its singletons are closed. T_1 spaces are always T_0.
- Sober. A space is sober if every irreducible closed set C has a unique generic point p. In other words, if C is not the (possibly nondisjoint) union of two smaller closed subsets, then there is a p such that the closure of $\{p\}$ equals C, and p is the only point with this property.
- T_2 or Hausdorff. A space is Hausdorff if every two distinct points have disjoint neighbourhoods. T_2 spaces are always T_1.
- $T_{2½}$ or Urysohn. A space is Urysohn if every two distinct points have disjoint *closed* neighbourhoods. $T_{2½}$ spaces are always T_2.
- Completely T_2 or completely Hausdorff. A space is completely T_2 if every two distinct points are separated by a function. Every completely Hausdorff space is Urysohn.
- Regular. A space is regular if whenever C is a closed set and p is a point not in C, then C and p have disjoint neighbourhoods.
- T_3 or Regular Hausdorff. A space is regular Hausdorff if it is a regular T_0 space. (A regular space is Hausdorff if and only if it is T_0, so the terminology is consistent.)
- Completely regular. A space is completely regular if whenever C is a closed set and p is a point not in C, then C and $\{p\}$ are separated by a function.
- $T_{3½}$, Tychonoff, Completely regular Hausdorff or Completely T_3. A Tychonoff space is a completely regular T_0 space. (A completely regular space is Hausdorff if and only if it is T_0, so the terminology is consistent.) Tychonoff spaces are always regular Hausdorff.
- Normal. A space is normal if any two disjoint closed sets have disjoint neighbourhoods. Normal spaces admit partitions of unity.
- T_4 or Normal Hausdorff. A normal space is Hausdorff if and only if it is T_1. Normal Hausdorff spaces are always Tychonoff.
- Completely normal. A space is completely normal if any two separated sets have disjoint neighbourhoods.

- T_5 or Completely normal Hausdorff. A completely normal space is Hausdorff if and only if it is T_1. Completely normal Hausdorff spaces are always normal Hausdorff.
- Perfectly normal. A space is perfectly normal if any two disjoint closed sets are precisely separated by a function. A perfectly normal space must also be completely normal.
- Perfectly normal Hausdorff, or perfectly T_4. A space is perfectly normal Hausdorff, if it is both perfectly normal and T_1. A perfectly normal Hausdorff space must also be completely normal Hausdorff.
- Discrete space. A space is discrete if all of its points are completely isolated, i.e. if any subset is open.

Countability Conditions

- Separable. A space is separable if it has a countable dense subset.
- Lindelöf. A space is Lindelöf if every open cover has a countable subcover.
- First-countable. A space is first-countable if every point has a countable local base.
- Second-countable. A space is second-countable if it has a countable base for its topology. Second-countable spaces are always separable, first-countable and Lindelöf.

Connectedness

- Connected. A space is connected if it is not the union of a pair of disjoint non-empty open sets. Equivalently, a space is connected if the only clopen sets are the empty set and itself.
- Locally connected. A space is locally connected if every point has a local base consisting of connected sets.
- Totally disconnected. A space is totally disconnected if it has no connected subset with more than one point.
- Path-connected. A space X is path-connected if for every two points x, y in X, there is a path p from x to y, i.e., a continuous map p: $[0,1] \to X$ with $p(0) = x$ and $p(1) = y$. Path-connected spaces are always connected.
- Locally path-connected. A space is locally path-connected if every point has a local base consisting of path-connected sets. A locally path-connected space is connected if and only if it is path-connected.

- Simply connected. A space X is simply connected if it is path-connected and every continuous map f: $S^1 \rightarrow X$ is homotopic to a constant map.
- Locally simply connected. A space X is locally simply connected if every point x in X has a local base of neighborhoods U that is simply connected.
- Semi-locally simply connected. A space X is semi-locally simply connected if every point has a local base of neighborhoods U such that *every* loop in U is contractible in X. Semi-local simple connectivity, a strictly weaker condition than local simple connectivity, is a necessary condition for the existence of a universal cover.
- Contractible. A space X is contractible if the identity map on X is homotopic to a constant map. Contractible spaces are always simply connected.
- Hyper-connected. A space is hyper-connected if no two non-empty open sets are disjoint. Every hyper-connected space is connected.
- Ultra-connected. A space is ultra-connected if no two non-empty closed sets are disjoint. Every ultra-connected space is path-connected.
- Indiscrete or trivial. A space is indiscrete if the only open sets are the empty set and itself. Such a space is said to have the trivial topology.

Compactness

- Compact. A space is compact if every open cover has a finite subcover. Some authors call these spaces quasicompact and reserve compact for Hausdorff spaces where every open cover has finite subcover. Compact spaces are always Lindelöf and paracompact. Compact Hausdorff spaces are therefore normal.
- Sequentially compact. A space is sequentially compact if every sequence has a convergent subsequence.
- Countably compact. A space is countably compact if every countable open cover has a finite subcover.
- Pseudocompact. A space is pseudocompact if every continuous real-valued function on the space is bounded.
- σ-compact. A space is σ-compact if it is the union of countably many compact subsets.
- Paracompact. A space is paracompact if every open cover has an open locally finite refinement. Paracompact Hausdorff spaces are normal.
- Locally compact. A space is locally compact if every point has a local base con-

sisting of compact neighbourhoods. Slightly different definitions are also used. Locally compact Hausdorff spaces are always Tychonoff.

- Ultraconnected compact. In an ultra-connected compact space *X* every open cover must contain *X* itself. Non-empty ultra-connected compact spaces have a largest proper open subset called a monolith.

Metrizability

- Metrizable. A space is metrizable if it is homeomorphic to a metric space. Metrizable spaces are always Hausdorff and paracompact (and hence normal and Tychonoff), and first-countable.Moreover a topological space (X,T) is said to be metrizable if there exists a metric for X such that the metric topology T(d) is identical with the topology T.
- Polish. A space is called Polish if it is metrizable with a separable and complete metric.
- Locally metrizable. A space is locally metrizable if every point has a metrizable neighbourhood.

Miscellaneous

- Baire space. A space *X* is a Baire space if it is not meagre in itself. Equivalently, *X* is a Baire space if the intersection of countably many dense open sets is dense.
- Topological Homogeneity. A space *X* is (topologically) homogeneous if for every x and y in *X* there is a homeomorphism $f: X \to X$ such that $f(x) = y$. Intuitively speaking, this means that the space looks the same at every point. All topological groups are homogeneous.
- Finitely generated or Alexandrov. A space *X* is Alexandrov if arbitrary intersections of open sets in *X* are open, or equivalently if arbitrary unions of closed sets are closed. These are precisely the finitely generated members of the category of topological spaces and continuous maps.
- Zero-dimensional. A space is zero-dimensional if it has a base of clopen sets. These are precisely the spaces with a small inductive dimension of 0.
- Almost discrete. A space is almost discrete if every open set is closed (hence clopen). The almost discrete spaces are precisely the finitely generated zero-dimensional spaces.
- Boolean. A space is Boolean if it is zero-dimensional, compact and Hausdorff (equivalently, totally disconnected, compact and Hausdorff). These are precisely the spaces that are homeomorphic to the Stone spaces of Boolean algebras.

- Reidemeister torsion
- κ-resolvable. A space is said to be κ-resolvable (respectively: almost κ-resolvable) if it contains κ dense sets that are pairwise disjoint (respectively: almost disjoint over the ideal of nowhere dense subsets). If the space is not κ-resolvable then it is called κ-irresolvable.
- Maximally resolvable. Space X is maximally resolvable if it is $\Delta(X)$-resolvable, where $\Delta(X) = \min\{|G| : G \neq \varnothing, G \text{ is open}\}$. Number $\Delta(X)$ is called dispersion character of X.
- Strongly discrete. Set D is strongly discrete subset of the space X if the points in D may be separated by pairwise disjoint neighborhoods. Space X is said to be strongly discrete if every non-isolated point of X is the accumulation point of some strongly discrete set.

References

- Arkhangel'skii, A.V. (2001), "Compact space", in Hazewinkel, Michiel, Encyclopedia of Mathematics, Springer, ISBN 978-1-55608-010-4
- Fréchet, Maurice (1906), "Sur quelques points du calcul fonctionnel", Rendiconti del Circolo Matematico di Palermo, 22 (1): 1–72, doi:10.1007/BF03018603
- Kline, Morris (1972), Mathematical thought from ancient to modern times (3rd ed.), Oxford University Press (published 1990), ISBN 978-0-19-506136-9
- Scarborough, C.T.; Stone, A.H. (1966), "Products of nearly compact spaces", Transactions of the American Mathematical Society, Transactions of the American Mathematical Society, Vol. 124, No. 1, 124 (1): 131–147, JSTOR 1994440, doi:10.2307/1994440
- Steen, Lynn Arthur; Seebach, J. Arthur Jr. (1995) [1978], Counterexamples in Topology (Dover Publications reprint of 1978 ed.), Berlin, New York: Springer-Verlag, ISBN 978-0-486-68735-3, MR 507446
- Barble M R Stadler; et al. "The Topology of the Possible: Formal Spaces Underlying Patterns of Evolutionary Change". Journal of Theoretical Biology. 213: 241–274. PMID 11894994. doi:10.1006/jtbi.2001.2423
- Gunnar Carlsson (April 2009). "Topology and data" (PDF). BULLETIN (New Series) OF THE AMERICAN MATHEMATICAL SOCIETY. 46 (2): 255–308. doi:10.1090/S0273-0979-09-01249-X
- Neuwirth, L. P., ed. (1975) Knots, Groups, and 3-Manifolds. Papers Dedicated to the Memory of R. H. Fox. Princeton University Press. ISBN 978-0-691-08170-0
- Cambou, Anne Dominique; Narayanan, Menon (2011). "Three-dimensional structure of a sheet crumpled into a ball.". Proceedings of the National Academy of Sciences. 108 (36): 14741–14745 doi:10.1073/pnas.1019192108
- Gowers, Timothy; Barrow-Green, June; Leader, Imre, eds. (2008), The Princeton Companion to Mathematics, Princeton University Press, ISBN 978-0-691-11880-2

2

Construction of Topological Spaces

A group of topological spaces containing a natural topology is known as product topology. Box topology and quotient space are other categories that are discussed here. This chapter elucidates the crucial theories and principles of topology.

Product Topology

In topology and related areas of mathematics, a product space is the cartesian product of a family of topological spaces equipped with a natural topology called the product topology. This topology differs from another, perhaps more obvious, topology called the box topology, which can also be given to a product space and which agrees with the product topology when the product is over only finitely many spaces. However, the product topology is "correct" in that it makes the product space a categorical product of its factors, whereas the box topology is too fine; this is the sense in which the product topology is "natural".

Definition

Given X such that

$$X := \prod_{i \in I} X_i$$

is the Cartesian product of the topological spaces X_i, indexed by $i \in I$, and the canonical projections $p_i : X \to X_i$, the product topology on X is defined to be the coarsest topology (i.e. the topology with the fewest open sets) for which all the projections p_i are continuous. The product topology is sometimes called the Tychonoff topology.

The open sets in the product topology are unions (finite or infinite) of sets of the form $\prod_{i \in I} U_i$, where each U_i is open in X_i and $U_i \neq X_i$ for only finitely many i. In particular, for a finite product (in particular, for the product of two topological spaces), the products of base elements of the X_i gives a basis for the product $\prod_{i \in I} X_i$.

The product topology on X is the topology generated by sets of the form $p_i^{-1}(U_i)$, where i is in I and U_i is an open subset of X_i. In other words, the sets $\{p_i^{-1}(U_i)\}$ form a subbase for the topology on X. A subset of X is open if and only if it is a (possibly infinite) union of intersections of finitely many sets of the form $p_i^{-1}(U_i)$. The $p_i^{-1}(U_i)$ are sometimes called open cylinders, and their intersections are cylinder sets.

In general, the product of the topologies of each X_i forms a basis for what is called the box topology on X. In general, the box topology is finer than the product topology, but for finite products they coincide.

Examples

If one starts with the standard topology on the real line R and defines a topology on the product of n copies of R in this fashion, one obtains the ordinary Euclidean topology on R^n.

The Cantor set is homeomorphic to the product of countably many copies of the discrete space {0,1} and the space of irrational numbers is homeomorphic to the product of countably many copies of the natural numbers, where again each copy carries the discrete topology.

Properties

The product space X, together with the canonical projections, can be characterized by the following universal property: If Y is a topological space, and for every i in I, $f_i : Y \to X_i$ is a continuous map, then there exists *precisely one* continuous map $f : Y \to X$ such that for each i in I the following diagram commutes:

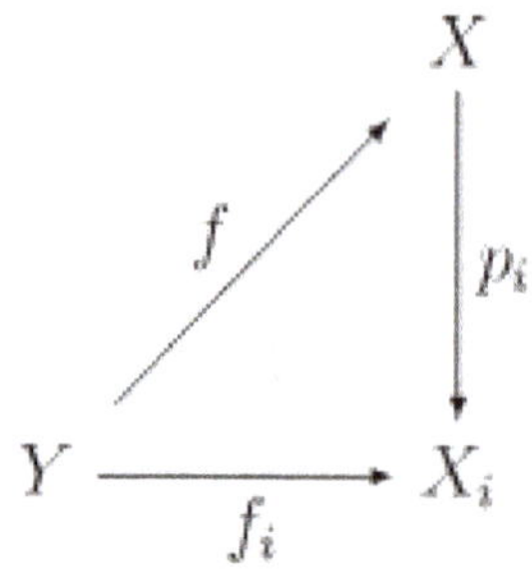

This shows that the product space is a product in the category of topological spaces. It follows from the above universal property that a map $f : Y \to X$ is continuous if and only if $f_i = p_i \circ f$ is continuous for all i in I. In many cases it is easier to check that the component functions f_i are continuous. Checking whether a map $f : Y \to X$ is continuous is usually more difficult; one tries to use the fact that the p_i are continuous in some way.

In addition to being continuous, the canonical projections $p_i : X \to X_i$ are open maps. This means that any open subset of the product space remains open when projected down to the X_i. The converse is not true: if W is a subspace of the product space whose projections down to all the X_i are open, then W need not be open in X. (Consider for instance $W = R^2 \setminus (0,1)^2$.) The canonical projections are not generally closed maps (consider for example the closed set $\{(x, y) \in \mathbb{R}^2 \mid xy = 1\}$ whose projections onto both axes are R \ {o}).

The product topology is also called the *topology of pointwise convergence* because of the following fact: a sequence (or net) in X converges if and only if all its projections to the spaces X_i converge. In particular, if one considers the space $X = R^I$ of all real valued functions on I, convergence in the product topology is the same as pointwise convergence of functions.

Any product of closed subsets of X_i is a closed set in X.

An important theorem about the product topology is Tychonoff's theorem: any product of compact spaces is compact. This is easy to show for finite products, while the general statement is equivalent to the axiom of choice.

Relation to other Topological Notions

- Separation
  - Every product of T_0 spaces is T_0
  - Every product of T_1 spaces is T_1
  - Every product of Hausdorff spaces is Hausdorff
  - Every product of regular spaces is regular
  - Every product of Tychonoff spaces is Tychonoff
  - A product of normal spaces *need not* be normal
- Compactness
  - Every product of compact spaces is compact (Tychonoff's theorem)
  - A product of locally compact spaces *need not* be locally compact. However, an arbitrary product of locally compact spaces where all but finitely many are compact *is* locally compact (This condition is sufficient and necessary).
- Connectedness
  - Every product of connected (resp. path-connected) spaces is connected (resp. path-connected)
  - Every product of hereditarily disconnected spaces is hereditarily disconnected.
- Metric spaces
  - Countable products of metric spaces are metrizable

Axiom of Choice

The axiom of choice is equivalent to the statement that the product of a collection of non-empty sets is non-empty. The proof is easy enough: one needs only to pick an element from each set to find a representative in the product. Conversely, a representative of the product is a set which contains exactly one element from each component.

The axiom of choice occurs again in the study of (topological) product spaces; for example, Tychonoff's theorem on compact sets is a more complex and subtle example of a statement that is equivalent to the axiom of choice.

Product Space

Definition. For each $\alpha \in J$, define a function $p_\alpha : X \to X_\alpha$, known as α-th projection or coordinate function, as $p_\alpha(f) = f(\alpha) = x_\alpha$.

Our aim is to define a topology $\mathcal{J}$ on $\prod_{\alpha\in J} X_\alpha$ which will have the following properties:

- Each projection function $p_\alpha : (X, \mathcal{J}) \to (X_\alpha, \mathcal{J}_\alpha)$ is continuous.
- Whenever $\mathcal{J}'$ is a topology on X such that each $p_\alpha : (X, \mathcal{J}') \to (X_\alpha, \mathcal{J}_\alpha)$ is continuous then $\mathcal{J} \subseteq \mathcal{J}'$.

That is $\mathcal{J}$ is the smallest (or weakest topology) on X that makes each p_α continuous. For $\alpha \in J$, let $S_\alpha = \{p_\alpha^{-1}(U_\alpha) : U_\alpha \in \mathcal{J}_\alpha\}$. Then we require that $p_\alpha^{-1}(U_\alpha)$ is open in our proposed topological space $(X, \mathcal{J})$. Hence we require that $\bigcup_{\alpha\in J} S_\alpha \subseteq \mathcal{J}$. Note that $p_\alpha^{-1}(U_\alpha) = \{f \in X : p_\alpha(f) = f(\alpha) = x_\alpha \in X_\alpha\}$. Hence if we fix an $\alpha \in J$, then $p_\alpha^{-1}(U_\alpha) = \prod_{\beta\in J} A_\beta$, where $A_\alpha = U_\alpha$ and $A_\beta = X_\beta$ when $\beta \neq \alpha$. If $\alpha_1, \alpha_2, \ldots, \alpha_n, n \in \mathbb{N}$ and $U_{\alpha_i} \in \mathcal{J}_{\alpha_i}$, $i = 1, 2, \ldots, n$, then $p_{\alpha_i}^{-1}(U_{\alpha_i}) \in \mathcal{J}$. Also $\mathcal{J}$ is closed under finite intersections means that $\bigcap_{i=1}^{n} p_{\alpha_i}^{-1}(U_{\alpha_i}) = \prod_{\alpha\in J} A_\alpha \in \mathcal{J}$, where $A_{\alpha_i} = U_{\alpha_i}, i = 1, 2, \ldots, n$ and $A_\alpha = X_\alpha$ when $\alpha \neq \alpha_1, \alpha_2, \ldots, \alpha_n$. Now it is easy to see that $\mathcal{B} = \{\prod_{\alpha\in J} U_\alpha : U_\alpha \in \mathcal{J}_\alpha$ for all $\alpha \in J$ *and* $U_\alpha = X_\alpha$, *except for finitely many* $\alpha \neq \alpha_1, \alpha_2, \ldots, \alpha_n \in J\}$ is a basis for a topology on X. The topology $\mathcal{J}$ induced by $\mathcal{B}$ is called the product topology on $X = \prod_{\alpha\in J} X_\alpha$ and the topological space $(X, \mathcal{J})$ is called the product topological space (also known as product space) induced by the topological spaces $(X_\alpha, \mathcal{J}_\alpha)$, $\alpha \in J$.

Remark: What will happen when J = {1, 2,...,n} for some natural number n?

When n=1, $n = 1$, $X = \{f : \{1\} \to X_1 : f(1) = x_1 \in X_1\} = X_1$ and $X = \{f : \{1, 2, \ldots, n\} \to \bigcup_{i=1}^{n} X_i : f(i) = x_i \in X_i\}$. That is $f = (f(1), f(2), \ldots, f(n)) = (x_1, x_2, \ldots, x_n) \in X = \prod_{i=1}^{n} X_i$.

Hence $X=\{(x_1, x_2, \ldots, x_n): x_i \in X_i, i=1, 2, \ldots, n\} = X_1 \times X_2 \times \cdots \times X_n$. In this case, that is when J = {1, 2, . . . , n} is a finite index set containing n elements, $\mathscr{B} = \left\{\prod_{i=1}^{n} U_i : \text{ each } U_i \text{ is open in } X_i, i=1,2,\ldots,n\right\}$ is a basis for the product topology $\mathcal{J}$ on $X=\prod_{i=1}^{n} X_i$. Now let us prove the following theorem:

Theorem 1. Let $J \neq \phi$ be an index set and $(X_\alpha, \mathcal{J}_\alpha)$, $\alpha \in J$ be a collection of Hausdorff topological spaces. Then the product space $\left(\prod_{\alpha\in J} X_\alpha, \mathcal{J}\right)$ is also a Hausdorff topological space.

Proof: Our aim is to prove that the product topological space $\left(\prod_{\alpha\in J} X_\alpha, \mathcal{J}\right)$ is a Haus dorff topological space. So take two distinct elements f, g in $\prod_{\alpha\in J} X_\alpha$. Now $f, g \in \prod_{\alpha\in J} X_\alpha$ implies $f: J \to \bigcup_{\alpha\in J} X_\alpha$ and $g: J \to \bigcup_{\alpha\in J} X_\alpha$ such that $f(\alpha) = x_\alpha \in X_\alpha$, $g(\alpha) = y_\alpha \in X_\alpha$, for each α ∈ J. Also $f \neq g \Rightarrow$ there exists $\alpha_0 \in J$ such that $x_{\alpha_0} = f(\alpha_0) \neq g(\alpha_0) = y_{\alpha_0}$. We have $x_{\alpha_0}, y_{\alpha_0} \in X_{\alpha_0}$ and $x_{\alpha_0} \neq y_{\alpha_0}$. Hence $(x_{\alpha_0}, \mathcal{J}_{\alpha_0})$ is a Hausdorff topological space implies that there exist $U_{\alpha_0}, V_{\alpha_0} \in \mathcal{J}_{\alpha_0}$ satisfying:

(i) $x_{\alpha_0} \in U_{\alpha_0}$, $y_{\alpha_0} \in V_{\alpha_0}$ and (ii) $U_{\alpha_0} \cap V_{\alpha_0} = \phi$.

Now use (i) and (ii) to construct basic open sets U, V in the product space satisfying f ∈ U, g ∈ V, and $U \cap V = \phi$. So, let $U_\alpha = X_\alpha$, $V_\alpha = X_\alpha$, whenever $\alpha \neq \alpha_0$. We already have $U_{\alpha_0}, V_{\alpha_0}$ which are open sets in $(X_{\alpha_0}, \mathcal{J}_{\alpha_0})$. Define U, V as $U = \prod_{\alpha\in J} U_\alpha$, $V = \prod_{\alpha\in J} V_\alpha$, where U, V are defined as above. We have *f(α)* ∈ X_α, *g(α)* ∈ X_α for all α ∈ *J* and hence *f* ∈ *U*, *g* ∈ *V* (why?). Also $U \cap V = \left(\prod_{\alpha\in J} U_\alpha\right) \cap \left(\prod_{\alpha\in J} V_\alpha\right) = \prod_{\alpha\in J} U_\alpha \cap V_\alpha = \phi$, *since* $U_{\alpha_0} \cap V_{\alpha_0} = \phi$. That is for $f, g \in \prod_{\alpha\in J} X_\alpha$ with $f \neq g$ there exist basic open sets U, V in the product space such that *f* ∈ *U*, *g* ∈ *V*, and $U \cap V = \phi$. This implies that the product space $\left(\prod_{\alpha\in J} X_\alpha, \mathcal{J}\right)$ is a Hausdorff space.

Note. Let $(X, \mathcal{J})$ be a topological space and $\mathscr{B}$ be a basis for $(X, \mathcal{J})$ (or say $\mathscr{B}$ is a basis for $\mathcal{J}$). Then for a subset A of X, $x \in \overline{A}$ if and only if for each $U \in \mathscr{B}$ with $x \in U, U \cap A \neq \phi$. That is $x \in \overline{A}$ if and only if for each basic open set U containing x, $U \cap A \neq \phi$.

Theorem 2. Let $(X_\alpha, \mathcal{J}_\alpha)$, $\alpha \in J$ be a collection of topological spaces and $A_\alpha \subseteq X_\alpha$ for each α ∈ *J* then $\prod_{\alpha\in J} \overline{A_\alpha} = \overline{\prod_{\alpha\in J} A_\alpha}$, with respect to the product space $\left(\prod_{\alpha\in J} X_\alpha, \mathcal{J}\right)$.

Proof: First let us prove $\prod_{\alpha\in J} \overline{A_\alpha} \subseteq \overline{\prod_{\alpha\in J} A_\alpha}$. Let $f \in \prod_{\alpha\in J} \overline{A_\alpha}$. Then $f: J \to \bigcup_{\alpha\in J} \overline{A_\alpha}$ such that $f(\alpha)=x_\alpha \in \overline{A}_\alpha$ for all α ∈ J. We aim to prove that f is in the closure of $\prod_{\alpha\in J} A_\alpha$ in the product topological space $\prod_{\alpha\in J} X_\alpha$. So take a basic open set B in the product space $\left(\prod_{\alpha\in J} X_\alpha, J\right)$ containing f, say $B=\prod_{\alpha\in J} U_\alpha$. It is given that $f \in \prod_{\alpha\in J} \overline{A}_\alpha$. Hence $f(\alpha)=x_\alpha \in \overline{A}_\alpha$ for each α ∈ J. Now $f \in B = \prod_{\alpha\in J} U_\alpha$ implies *f(α)* ∈ U_α for all α ∈ J. That is U_α is an open set containing x_α and $x_\alpha \in \overline{A}_\alpha$. This implies that $U_\alpha \cap A_\alpha \neq \phi$. Let $z_\alpha \in U_\alpha \cap A_\alpha$ for all α ∈ J. Define $g : J \to \bigcup_{\alpha\in J} A_\alpha$ as g(α) = $z_\alpha \in A_\alpha$ then $g \in B \cap \prod_{\alpha\in J} A_\alpha$. This implies that for each basic open set B containing $f, B \cap \prod_{\alpha\in J} A_\alpha \neq \phi$. This implies $f \in \overline{\prod_{\alpha\in J} A_\alpha}$. That is $f \in \prod_{\alpha\in J} \overline{A_\alpha}$ implies $f \in \overline{\prod_{\alpha\in J} A_\alpha}$ and this proves the assertion

$$\prod_{\alpha\in J} \overline{A}_\alpha \subseteq \overline{\prod_{\alpha\in J} A_\alpha} \qquad (1)$$

Now let us prove the converse part namely $\overline{\prod_{\alpha\in J} A_\alpha} \subseteq \prod_{\alpha\in J} \overline{A}_\alpha$. So let $f \in \overline{\prod_{\alpha\in J} A_\alpha} \subseteq \prod_{\alpha\in J} X_\alpha = X$. Our aim is to prove: $f \in \prod_{\alpha\in J} \overline{A}_\alpha$. But $f \in \prod_{\alpha\in J} X_\alpha$ if and only if *f(α)* ∈ X_α for each α ∈ J and $f \in \prod_{\alpha\in J} \overline{A}_\alpha$ if and only if $f(\alpha) \in \overline{A}_\alpha$ for each α ∈ J. For a fixed $\alpha_0 \in J$ take an open set U_{α_0} containing $f(\alpha_0)=x_{\alpha_0}$. We will have to use the fact that $f \in \overline{\prod_{\alpha\in J} A_\alpha}$. To use this fact we will have to construct a basic open set containing f. Keeping this in mind, we define $B=\prod_{\alpha\in J} U_\alpha$, where $U_\alpha = X_\alpha$, when $\alpha \neq \alpha_0$ and U_{α_0} is as given above. Now this B is a basic open set containing f and hence $f \in \overline{\prod_{\alpha\in J} A_\alpha}$ implies $B \cap \prod_{\alpha\in J} A_\alpha \neq \phi$. That is $\left(\prod_{\alpha\in J} U_\alpha\right) \cap \left(\prod_{\alpha\in J} A_\alpha\right) = \prod_{\alpha\in J} U_\alpha \cap A_\alpha \neq \phi$. This implies that each $U_\alpha \cap A_\alpha \neq \phi$. In particular $U_{\alpha_0} \cap A_{\alpha_0} \neq \phi$ implies $x_{\alpha_0} \in \overline{A}_{\alpha_0}$. Note that though our $\alpha_0 \in J$ is a fixed element, there is no restriction on $\alpha_0 \in J$ and the proof will go through for any α ∈ J. This gives that $f(\alpha)=x_\alpha \in \overline{A}_\alpha$ for all α ∈ J and this implies that $f \in \prod_{\alpha\in J} \overline{A}_\alpha$. Hence $f \in \overline{\prod_{\alpha\in J} A_\alpha} \Rightarrow f \in \prod_{\alpha\in J} \overline{A}_\alpha$. This implies

$$\overline{\prod_{\alpha\in J} A_\alpha} \subseteq \prod_{\alpha\in J} \overline{A}_\alpha \qquad (2)$$

Now combining Eqs. (1) and (2) we have $\prod_{\alpha\in J} \overline{A}_\alpha = \overline{\prod_{\alpha\in J} A_\alpha}$.

Box Topology

In topology, the cartesian product of topological spaces can be given several different topologies. One of the more obvious choices is the box topology, where a base is given by the Cartesian products of open sets in the component spaces. Another possibility is the product topology, where a base is given by the Cartesian products of open sets in the component spaces, only finitely many of which can be not equal to the entire component space.

While the box topology has a somewhat more intuitive definition than the product topology, it satisfies fewer desirable properties. In particular, if all the component spaces are compact, the box topology on their Cartesian product will not necessarily be compact, although the product topology on their Cartesian product will always be compact. In general, the box topology is finer than the product topology, although the two agree in the case of finite direct products (or when all but finitely many of the factors are trivial).

Definition

Given X such that

$$X := \prod_{i \in I} X_i,$$

or the (possibly infinite) Cartesian product of the topological spaces X_i, indexed by $i \in I$, the box topology on X is generated by the base

$$B = \left\{ \prod_{i \in I} U_i \,\middle|\, U_i \text{ open in } X_i \right\}.$$

The name *box* comes from the case of R^n, the basis sets look like boxes or unions thereof.

Properties

Box topology on R^ω:

- The box topology is completely regular
- The box topology is neither compact nor connected
- The box topology is not first countable (hence not metrizable)
- The box topology is not separable
- The box topology is paracompact (and hence normal and completely regular) if the continuum hypothesis is true

Example - Failure at Continuity

The following example is based on the Hilbert cube. Let R^ω denote the countable cartesian product of R with itself, i.e. the set of all sequences in R. Equip R with the standard topology and R^ω with the box topology. Let $f : R \to R^\omega$ be the product map whose components are all the identity, i.e. $f(x) = (x, x, x, \ldots)$. Although each component function is continuous, f is not continuous. To see this, consider the open set $U = \prod_{n=1}^{\infty} \left(-\frac{1}{n}, \frac{1}{n}\right)$. Since $f(0) = (0, 0, 0, \ldots) \in U$, if f were continuous, then there would exist some $\varepsilon > 0$ such that $(-\varepsilon, \varepsilon) \subseteq f^{-1}(U)$. But this would imply that $f(\varepsilon/2) = (\varepsilon/2, \varepsilon/2, \varepsilon/2, \ldots) \in U$ which is false since $\varepsilon/2 > 1/n$ for $n > \lceil 2/\varepsilon \rceil$. Thus f is not continuous even though all its component functions are.

Example - Failure at Compactness

Consider the countable product $X = \prod X_i$ where for each i, $X_i = \{0,1\}$ with the discrete topology. The box topology on X will also be the discrete topology. Consider the sequence $\{x_n\}_{n=1}^{\infty}$ given by

$$(x_n)_m = \begin{cases} 0 & m < n \\ 1 & m \geq n \end{cases}$$

Since no two points in the sequence are the same, the sequence has no limit point, and therefore X is not compact, even though its component spaces are.

Intuitive Description of Convergence; Comparisons

Topologies are often best understood by describing how sequences converge. In general, a cartesian product of a space X with itself over an indexing set S is precisely the space of functions from S to X; the product topology yields the topology of pointwise convergence; sequences of functions converge if and only if they converge at every point of S. The box topology, once again due to its great profusion of open sets, makes convergence very hard. One way to visualize the convergence in this topology is to think of functions from R to R—a sequence of functions converges to a function f in the box topology if, when looking at the graph of f, given any set of "hoops", that is, vertical open intervals surrounding the graph of f above every point on the x-axis, eventually, every function in the sequence "jumps through all the hoops." For functions on R this looks a lot like uniform convergence, in which case all the "hoops", once chosen, must be the same size. But in this case one can make the hoops arbitrarily small, so one can see intuitively how "hard" it is for sequences of functions to converge. The hoop picture works for convergence in the product topology as well: here we only require all the functions to jump through any given *finite* set of hoops. This stems directly from the fact that, in the product topology, almost all the factors in a basic open set are the whole space. Interestingly, this is actually equivalent to requiring all functions to eventually jump through just a *single* given hoop; this is just the definition of pointwise convergence.

Comparison with Product Topology

The basis sets in the product topology have almost the same definition as the above, *except* with the qualification that *all but finitely many* U_i are equal to the component space X_i. The product topology satisfies a very desirable property for maps $f_i : Y \to X_i$ into the component spaces: the product map $f: Y \to X$ defined by the component functions f is continuous if and only if all the f_i are continuous. As shown above, this does not always hold in the box topology. This actually makes the box topology very useful for providing counterexamples—many qualities such as compactness, connectedness, metrizability, etc., if possessed by the factor spaces, are not in general preserved in the product with this topology.

The Box Topology

Definition: Let $(X_\alpha, \mathcal{J}_\alpha), \alpha \in J$, be a collection of topological spaces and $\mathcal{B}_b = \{\prod_{\alpha \in J} U_\alpha : U_\alpha \in \mathcal{J}_\alpha \text{ for } \alpha \in J\}$. Then $\mathcal{B}_b$ is a basis for a topology on $X = \prod_{\alpha \in J} X_\alpha$ and $\mathcal{J}_b$, the topology induced by $\mathcal{B}_b$, is called the box topology on X.

Remark: From the definitions of product and box topologies, it is clear that $\mathcal{B} \subseteq \mathcal{B}_b$, where the product topology $\mathcal{J}$ on X is induced by $\mathcal{B}$ (refer the definition of product topology) and the box topology $\mathcal{J}_b$ is induced by $\mathcal{B}_b$. Now $\mathcal{B} \subseteq \mathcal{B}_b$ implies $\mathcal{J}_{\mathcal{B}} = \mathcal{J} \subseteq \mathcal{J}_b$. That is the product topology on $\prod_{\alpha \in J} X_\alpha = X$ is weaker than the box topology $\mathcal{J}_b$ on X.

It is to be noted that if a subset A of X is open with respect to the product topology on X then it is also open with respect to the box topology on X. Note that the set $A = \prod_{i=1}^{\infty} \left(\frac{-1}{n}, \frac{1}{n} \right)$ is an open set in $\mathbb{R}^w = \mathbb{R} \times \mathbb{R} \times \cdots$ with respect to the box topology but not open with respect to the product topology on $\mathbb{R}^w$. Also we have proved that if $(X_\alpha, \mathcal{J}_\alpha), \alpha \in J$, is a collection of Hausdorff topological spaces then the product space $\left(\prod_{\alpha \in J} X_\alpha, \mathcal{J}_\alpha \right)$ is also a Hausdorff space. Since $\mathcal{J} \subseteq \mathcal{J}_b$ it is clear that if $(X_\alpha, \mathcal{J}_\alpha), \alpha \in J$ is a collection of Hausdorff topological spaces then $\left(\prod_{\alpha \in J} X_\alpha, \mathcal{J}_b \right)$ is also a Hausdorff space.

Note that if J is a nonempty finite index set then $\mathcal{J} = \mathcal{J}_b$ on $\prod_{\alpha \in J} X_\alpha$.

Theorem 3. Let $(X_\alpha, \mathcal{J}_\alpha), \alpha \in J$, be a collection of topological spaces and for each $\alpha \in J$, let $A_\alpha \subseteq X_\alpha$. Then $\prod_{\alpha \in J} \overline{A}_\alpha = \overline{\prod_{\alpha \in J} A_\alpha}$, where $\overline{A}_\alpha$ denotes the closure of A_α in $(X_\alpha, \mathcal{J}_\alpha)$ and $\overline{\prod_{\alpha \in J} A_\alpha}$ denotes the closure of $\prod_{\alpha \in J} A_\alpha$ in $\left(\prod_{\alpha \in J} X_\alpha, \mathcal{J}_b \right)$.

Theorem 4. Let $(X, \mathcal{J}), (Y, \mathcal{J}'), (Z, \mathcal{J}'')$ be topological spaces and

$f:(X, \mathcal{J}) \to (Y, \mathcal{J}')$, $g:(Y, \mathcal{J}') \to (Z, \mathcal{J}'')$ be continuous functions then the composite function $g \circ f:(X, \mathcal{J}) \to (Z, \mathcal{J}'')$ defined as $(g \circ f)(x) = g(f(x))$ is also a continuous function.

Proof: We aim to prove $g \circ f:(X, \mathcal{J}) \to (Z, \mathcal{J}'')$ is a continuous function. So start with an open set W in $(Z, \mathcal{J}'')$. Now W is an open set in Z (means $W \in \mathcal{J}''$) and $g:(Y, \mathcal{J}') \to (Z, \mathcal{J}'')$ is a continuous function implies $g^{-1}(W)$ is an open set Y. Now $f:(X, \mathcal{J}) \to (Y, \mathcal{J}')$ is also a continuous function. Hence $f^{-1}(g^{-1}(W))$ is an open set in X. We define for $A \subseteq Y, f^{-1}(A) = \{x \in X : f(x) \in A\}$. That is $x \in f^{-1}(A)$ if and only if $f(x) \in A$. Hence $f^{-1}(g^{-1}(W)) = \{x \in X : f(x) \in g^{-1}(W)\} = \{x \in X : g(f(x)) \in W\} = (g \circ f)^{-1}(W)$. That is we have proved: W is an open set in $(Z, \mathcal{J}'')$ implies $(g \circ f)^{-1}(W)$ is an open in $(X, \mathcal{J})$ implies $g \circ f:(X, \mathcal{J}) \to (Z, \mathcal{J}'')$ is a continuous function.

Definition: A sequence $\{x_n\}$ is a topological space $(X, \mathcal{J})$ is said to converge an element x in X if for each open set U containing x, there exists a natural number n_0 (that is $n_0 \in \mathbb{N}$) such that $x_n \in U$ for all $n \geq n_0$.

The product topology $\mathbb{R} \times \mathbb{R} \times \mathbb{R} \times \cdots = \prod_{n=1}^{\infty} \mathbb{R}_n = \mathbb{R}^w$, where $\mathbb{R}_n = \mathbb{R}$, $n = 1,2,3...$ is metrizable. That is, we will have to define a metric say d_1 of $\mathbb{R}^w$ such that $\mathcal{J}_{d_1} = \mathcal{J}$, the product topology on $\mathbb{R}^w$. For $x = (x_n)_{n=1}^{\infty} = (x_n) \in \mathbb{R}^w$, $y = (y_n) \in \mathbb{R}^w$, let $d_1(x,y) = \sup_{n \geq 1}\left\{\frac{\bar{d}(x_n, y_n)}{n}\right\}$, where $\bar{d}(x_n, y_n) = \min\{1, |x_n - y_n|\}$. (Exercise. Let (X, d) be a metric space and $\bar{d}(x, y) = \min\{1, |x - y|\}$ for all $x, y \in X$. Prove that (i) $\bar{d}$ is a metric on X, (ii) $\mathcal{J}_{\bar{d}} = \mathcal{J}_d$.)

It is easy to prove that d_1 is a metric on $\mathbb{R}^w$. First let us prove that $\mathcal{J}_{d_1} \subseteq \mathcal{J}$. So, let $U \in \mathcal{J}_{d_1}$. We aim to prove that each point of U is an interior point of U with respect to the product topology $\mathcal{J}$ on $\mathbb{R}^w$. Take $x \in U$. Now $x \in U$, U is an open set in the metric space $(\mathbb{R}^w, d_1)$ implies there exists r > o such that $B_{d_1}(x, r) \subseteq U$. Now choose $n_0 \in \mathbb{N}$ such that $\frac{1}{n_0} < r$ and $B = (x_1 - \epsilon, x_1 + \epsilon) \times \cdots \times (x_{n_0} - \epsilon, x_{n_0} + \epsilon) \times \mathbb{R} \times \mathbb{R} \times \cdots$ then B is a basic open set in $(\mathbb{R}^w, \mathcal{J})$ containing $x = (x_n)$. Now we leave it as an exercise to prove that $B \subseteq B_{d_1}(x, \epsilon)$. Hence for each $x \in U$, there exists a basic open set B in $(\mathbb{R}^w, \mathcal{J})$ such that $x \in B \subseteq U$. This proves that $U \in \mathcal{J}$ that is

$$\mathcal{J}_{d_1} \subseteq \mathcal{J}. \qquad (3)$$

Now let us prove that $\mathcal{J} \subseteq \mathcal{J}_{d_1}$. To prove this statement it is enough to prove that every basic open subset V of $(\mathbb{R}^w, \mathcal{J})$ is in $\mathcal{J}_{d_1}$. Now V is a basic open set in the product topology implies there exists $k \in \mathbb{N}$ such that $V = V_1 \times V_2 \times \cdots \times V_k \times \mathbb{R} \times \mathbb{R} \times \cdots$. Let $x = (x_n)_{n=1}^{\infty} \in V$. Hence there exist $\epsilon_1, \epsilon_2 \ldots \epsilon_k$, $0 < \epsilon_i < 1$ for $i = 1, 2, \ldots, k$ such that $(x_i - \epsilon_i, x_i + \epsilon_i) \subseteq V_i$. Now let $\epsilon = min\left\{\frac{\epsilon_i}{i} : i = 1, 2, \ldots, k\right\}$. (note: we have $U_i = \mathbb{R}$ for all $i > k$ and hence it is enough to consider $\epsilon_1, \ldots \epsilon_k$) and we claim that $B_{d_1}(x, \epsilon) \subseteq V$. So, let $y \in B_{d_1}(x, \epsilon)$ then $d_1(x, y) < \epsilon$ implies $\sup_{n \geq 1}\left\{\frac{\bar{d}(x_n, y_n)}{n}\right\} < \epsilon$ implies $\frac{\bar{d}(x_n, y_n)}{n} < \epsilon$ for all n=1,2,...,k implies $\frac{1}{n} min\{1, |x_n - y_n|\} < \epsilon$ for all n=1,2,...,k implies $min\{1, |x_n - y_n| < n\epsilon < \epsilon_n < 1$ for all n=1,2,...,k implies $|x_n - y_n| < \epsilon_n$ for all n=1,2,...,k implies $y = (y_n) \in V_1 \times \cdots \times V_k \times \mathbb{R} \times \cdots = V$

This proves that $B_{d_1}(x, \epsilon) \subseteq V$. That is for each $x \in V$ there exists $\epsilon > 0$ such that $B_{d_1}(x, \epsilon)$. Hence every point of V is an interior point of V with respect to $(\mathbb{R}^w, J_{d_1})$. Hence

$$V \in J_{d_1}. \tag{4}$$

Now if $U \in J_{d_1}$ then there exists $k \in \mathbb{N}$ and $B_1, B_2, \ldots, B_k \in \mathcal{B}$ ($\mathcal{B} = \{\prod_{n=1}^{\infty} U_n :$ each U_n is open in $\mathbb{R}$ and $U_n = \mathbb{R}$ for except finitely many n's} is our standard basis for the product topology J on $\mathbb{R}^w$) such that $U = B_1 \cap \cdots \cap B_k$. We have proved that each basic open set B of J belongs to J_{d_1} (i.e $B \in J_{d_1}$) (refer Eq. (4)). Now $B_1, B_2, \ldots, B_k \in J_{d_1}$ and J_{d_1} is a topology implies $B_1 \cap B_2 \cap \cdots \cap B_k \in J_{d_1}$. This proves that $U \in J_{d_1}$ That is

$$U \in J \Rightarrow U \in J_{d_1} \Rightarrow J \subseteq J_{d_1} \tag{5}$$

From Eqs. (3) and (5) we see that $J = J_{d_1}$. Hence the product space $\mathbb{R}^w = \mathbb{R} \times \mathbb{R} \times \cdots$ with product topology J is metrizable. It is interesting to note that if we consider the box topology say J_b on $\mathbb{R}^w$, then $(\mathbb{R}^w, J_b)$ is not a metrizable topological space.

How to prove that a given topological space (X, J) is not metrizable. If (X, J) is metrizable then we will have to find a metric (finding such a metric is not at all an easy task and this statement will become meaningful if we have patience to wait and see the proof of the Urysohn metrization theorem) say d on X such that $J_d = J$. We have just proved that $(\mathbb{R}^w, J)$ (with product topology) is metrizable and in this case we could define a metric d on $\mathbb{R}^w$ such that $J_d = J$.

To prove that a topological space is not metrizable space is comparatively easier. For example, if the given topological space (X, J) is not a Hausdorff topological space then it is clear that there cannot exist any metric d on X such that $\mathcal{J}_d = \mathcal{J}$. We know that if d is a metric on X, then $(X, \mathcal{J}_d)$ is a Hausdorff space. So, what we need here is to find a property a metric space has whereas the given topological space does not have that particular property. Now let us come back and prove that $(\mathbb{R}^w, \mathcal{J}_b)$ is not metrizable. Suppose there exists a metric say d on $\mathbb{R}^w$ such that $\mathcal{J}_d = \mathcal{J}_b$. Then we know that for $A \subseteq X$, $x \in \bar{A}$ if and only if there exists a sequence (x_n) in A such that $(x_n) \to x$ as $n \to \infty$ (to prove this statement, observe the following: For $x \in \bar{A}$, $B\left(x, \frac{1}{n}\right) \cap A \neq \phi$, for each $n \in \mathbb{N}$ and hence we have $\left\{B\left(x, \frac{1}{n}\right) \cap A\right\}_{n=1}^{\infty} = \{A_n\}_{n=1}^{\infty} = \{A_n\}_{n \in \mathbb{N}}$ a collection of nonempty sets. Now by axiom of choice there exists a choice function say $f : \mathbb{N} \to \bigcup_{n=1}^{\infty} A_n$ such that $x_n = f(n) \in A_n$. So using axiom of choice we have got a sequence (x_n) in A and now it is easy to see that $x_n \to x$ as $n \to \infty$).

Note that normally we just say that $B\left(x, \frac{1}{n}\right) \cap A \neq \phi$ implies there exists $x_n \in B\left(x, \frac{1}{n}\right) \cap A = A_n$. In such case it is to be understood that in fact we use axiom of choice to define such a sequence (x_n). Now let us prove that if $A = \{(x_1, x_2, x_3 \ldots,) : x_k > 0$ for all $k \in \mathbb{N}\}$. Then $0 = (0,0,0,\ldots,) \in \bar{A}$, but there does not exist any sequence $(x^{(n)})$ in A such that $x^{(n)} \to (0, 0,\ldots)$ in $(\mathbb{R}^w, \mathcal{J}_b)$.

Step 1: Prove that $0 \in \bar{A}$.

So take an open set say U containing o then there exists a basic open set $B = \prod_{n=1}^{\infty} B_n = B_1 \times B_2 \times \cdots$ such that $(0,0,\ldots,) \in B_1 \times B_2 \times \cdots \subseteq U$. (Here each B_k is an open set in $\mathbb{R}$ containing $0 \in \mathbb{R}$) $0 \in B_k$, $k = 1, 2, 3 \ldots$ implies there exist $a_k, b_k \in \mathbb{R}$, $a_k < b_k$ such that $0 \in (a_k, b_k) \subseteq B_k$, $b_k > 0$ and hence $\frac{b_k}{2} > 0$ implies $b = \left(\frac{b_k}{2}\right)_{k=1}^{\infty} \in A \cap B$ implies $A \cap B \neq \phi$ implies $A \cap U \neq \phi$ implies $0 = (0,0,\ldots,) \in \bar{A}$. Now we claim that there cannot exist any sequence $x^{(n)}$ in A such that $x^{(n)} \to (0, 0, 0 \ldots)$. Let $x^{(n)} = (a_{1n}, a_{2n}, \ldots,) \in A$. Then each $a_{in} > o$ for all $i = 1, 2,\ldots$ In particular, $a_{kk} > o$ for all $k = 1, 2, \ldots$. Let $U = \left(\frac{(-a_{11})}{2}, \frac{a_{11}}{2}\right) \times \left(\frac{(-a_{22})}{2}, \frac{a_{22}}{2}\right) \times \cdots$ then U is an open set in $(\mathbb{R}^w, \mathcal{J}_b)$ containing o.

What will happen if $U \cap A \neq \phi$. Note that for each n, $a_{nn} \notin \left(\left(\frac{-a_{nn}}{2}\right), \frac{a_{nn}}{2}\right)$ and hence $x^{(n)} = (a_{1n}, a_{2n}, \ldots,) \notin U$. If $x^{(n)} \to (0, 0, \ldots)$ then there exists $n_0 \in \mathbb{N}$ such that $x^{(n)} \in U$ for all $n \geq n_o$. But here $x^{(n)} \notin U$ for every n. Hence $x^{(n)}$ does not converge to (o,o,o,...). So $(0, 0, \ldots) \in \bar{A}$ but there cannot exist any sequence in A which converges to (0, 0,...) with respect to $\mathcal{J}_b$. This proves that $\left(\mathbb{R}^w, \mathcal{J}_b\right)$ is not a metrizable topological space.

Quotient Space (Topology)

In topology and related areas of mathematics, a quotient space (also called an identification space) is, intuitively speaking, the result of identifying or "gluing together" certain points of a given topological space. The points to be identified are specified by an equivalence relation. This is commonly done in order to construct new spaces from given ones. The quotient topology consists of all sets with an open preimage under the canonical projection map that maps each element to its equivalence class.

Definition

Let (X, τ_X) be a topological space, and let $\sim$ be an equivalence relation on X. The quotient space, $Y = X / \sim$ is defined to be the set of equivalence classes of elements of X:

$$Y = \left\{[x] : x \in X\right\} = \left\{\{v \in X : v \sim x\} : x \in X\right\},$$

equipped with the topology where the open sets are defined to be those sets of equivalence classes whose unions are open sets in X:

$$\tau_Y = \left\{U \subseteq Y : \bigcup U = \left(\bigcup_{[a]\in U} [a]\right) \in \tau_X\right\}.$$

Equivalently, we can define them to be those sets with an open preimage under the surjective map $q : X \to X / \sim$, which sends a point in X to the equivalence class containing it:

$$\tau_Y = \left\{U \subseteq Y : q^{-1}(U) \in \tau_X\right\}.$$

The quotient topology is the final topology on the quotient space with respect to the map q.

Quotient Map

A map $f : X \to Y$ is a quotient map (sometimes called an identification map) if it is surjective, and a subset U of Y is open if and only if $f^{-1}(U)$ is open. Equivalently, f is a quotient map if it is onto and Y is equipped with the final topology with respect to f.

Given an equivalence relation $\sim$ on X, the canonical map $q : X \to X/\sim$ is a quotient map.

Examples

- Gluing. Topologists talk of gluing points together. If X is a topological space and points $x, y \in X$ are to be "glued", then what is meant is that we are to consider the quotient space obtained from the equivalence relation $a \sim b$ if and only if $a = b$ or $a = x$, $b = y$ (or $a = y$, $b = x$).

- Consider the unit square $I^2 = [0,1] \times [0,1]$ and the equivalence relation ~ generated by the requirement that all boundary points be equivalent, thus identifying all boundary points to a single equivalence class. Then $I^2/\sim$ is homeomorphic to the unit sphere S^2.

- Adjunction space. More generally, suppose X is a space and A is a subspace of X. One can identify all points in A to a single equivalence class and leave points outside of A equivalent only to themselves. The resulting quotient space is denoted X/A. The 2-sphere is then homeomorphic to the unit disc with its boundary identified to a single point: $D^2 / \partial D^2$.

- Consider the set $X = \mathbb{R}$ of all real numbers with the ordinary topology, and write $x \sim y$ if and only if $x - y$ is an integer. Then the quotient space $X/\sim$ is homeomorphic to the unit circle S^1 via the homeomorphism which sends the equivalence class of x to $\exp(2\pi i x)$.

- A generalization of the previous example is the following: Suppose a topological group G acts continuously on a space X. One can form an equivalence relation on X by saying points are equivalent if and only if they lie in the same orbit. The quotient space under this relation is called the orbit space, denoted X/G. In the previous example $G = Z$ acts on R by translation. The orbit space R/Z is homeomorphic to S^1.

Note: The notation R/Z is somewhat ambiguous. If Z is understood to be a group acting on R then the quotient is the circle. However, if Z is thought of as a subspace of R, then the quotient is a countably infinite bouquet of circles joined at a single point.

Properties

Quotient maps $q : X \to Y$ are characterized among surjective maps by the following property: if Z is any topological space and $f : Y \to Z$ is any function, then f is continuous if and only if $f \circ q$ is continuous.

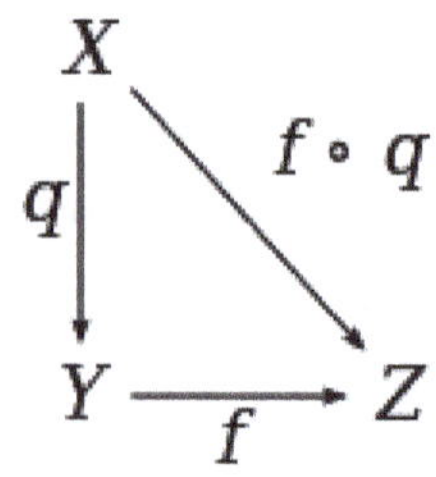

The quotient space $X/\sim$ together with the quotient map $q : X \to X/\sim$ is characterized by the following universal property: if $g : X \to Z$ is a continuous map such that $a \sim b$ implies $g(a) = g(b)$ for all a and b in X, then there exists a unique continuous map $f : X/\sim \to Z$ such that $g = f \circ q$. We say that g *descends to the quotient*.

The continuous maps defined on $X/\sim$ are therefore precisely those maps which arise from continuous maps defined on X that respect the equivalence relation (in the sense that they send equivalent elements to the same image). This criterion is copiously used when studying quotient spaces.

Given a continuous surjection $q : X \to Y$ it is useful to have criteria by which one can determine if q is a quotient map. Two sufficient criteria are that q be open or closed. Note that these conditions are only sufficient, not necessary. It is easy to construct examples of quotient maps that are neither open nor closed. For topological groups, the quotient map is open.

Compatibility with other Topological Notions

- Separation
    - In general, quotient spaces are ill-behaved with respect to separation axioms. The separation properties of X need not be inherited by $X/\sim$, and $X/\sim$ may have separation properties not shared by X.
    - $X/\sim$ is a T1 space if and only if every equivalence class of $\sim$ is closed in X.
    - If the quotient map is open, then $X/\sim$ is a Hausdorff space if and only if $\sim$ is a closed subset of the product space $X \times X$.
- Connectedness
    - If a space is connected or path connected, then so are all its quotient spaces.
    - A quotient space of a simply connected or contractible space need not share those properties.
- Compactness
    - If a space is compact, then so are all its quotient spaces.
    - A quotient space of a locally compact space need not be locally compact.
- Dimension
    - The topological dimension of a quotient space can be more (as well as less) than the dimension of the original space; space-filling curves provide such examples.

Quotient (Identification) Spaces

We start with a given topological space $(X, \mathcal{J})$. By identifying some of the points of X we can produce a new topology on a new set say X^*. For example if we consider the closed unit ball in $\mathbb{R}^2$, then our given topological space is $(X, \mathcal{J})$, where X is the closed unit ball in $\mathbb{R}^2$. Here we consider $(X, \mathcal{J})$ as a subspace of the Euclidean space $\mathbb{R}^2$. Now we get a new set $X^* = \{(x_1, x_2) \in \mathbb{R}^2 : x_1^2 + x_2^2 < 1\} \cup \{S'\}$, where S' is the unit circle (boundary) of the closed disc X. By defining a suitable topology $\mathcal{J}^*$ on X^* we can show that $(X^*, \mathcal{J}^*)$ is homeomorphic to the 2-sphere $S^2 = \{(x_1, x_2, x_3) \in \mathbb{R}^3 : x_1^2 + x_2^2 + x_3^2 = 1\}$. It is to be noted that here we are considering S^2 as a subspace of $\mathbb{R}^3$ (also note that if no topology on $\mathbb{R}^n$, $n \geq 1$ is mentioned then it is understood that we have the usual topology on $\mathbb{R}^n$).

Now let us see how to construct the quotient topology. Let $(X, \mathcal{J})$ be a topological space and X^* be a nonempty set. Let $p : X \to X^*$ be a surjective map.

Then $\mathcal{J}^* = \{A \subseteq Y : p^{-1}(A)$ is open in $(X, \mathcal{J})\}$ is a topology on X^*. This topology $\mathcal{J}^*$ on X^* is called the quotient topology on X^* induced by p.

It is easy to prove that $\mathcal{J}^*$ is a topology on X and we leave it as an exercise.

Definition: Let $(X, \mathcal{J})$ be a topological space and X^* be a partition of X into disjoint subsets whose union is X. Let $p : X \to X^*$ be the natural map satisfying the condition namely $x \in p(x)$, for each $x \in X$. Suppose for a given $x \in X$ there exist $A, B \in X^*$ such that $x \in A$ and $x \in B$. Then $x \in A \cap B$. This implies B = A. Hence for each $x \in X$ there exists a unique $A \in X^*$ such that $x \in A$ and this A is our $p(x)$. Also $\bigcup_{A \in X^*} A = X$ implies that p is onto. The quotient topology $\mathcal{J}^*$ on X^* is induced by p and we say that $(X^*, \mathcal{J}^*)$ is a quotient topology of $(X, \mathcal{J})$.

Let $(X, \mathcal{J})$ be a topological space and X^* be a partition of X into disjoint subsets whose union is X. Define a relation R on X as follows:

$R = \{(x, y) \in X \times X : x, y \in A$ for some $A \in X^*\}$ then (i) xRx, that is $(x, x) \in R$ for all $x \in X$, (ii) for $x, y \in X$, xRy implies there exists $A \in A^*$ such that $x, y \in A$. Hence $y, x \in A$ and this gives yRx that is for $x, y \in X$ $xRy \Rightarrow yRx$, (iii) for $x, y, z \in X$, xRy and yRz implies there exist $A, B \in X^*$ such that $x, y \in A$ and $y, z \in B$. Therefore $y \in A \cap B$ and this implies that $A = B$. From this we have $x, z \in A$. Hence xRz. That is xRy and yRz implies xRz. From (i), (ii) and (ii) we see that R is an equivalence relation on X and hence this relation R will partition X into disjoint equivalence classes.

For each $x \in X$, the equivalence classes determined by x is given by $\overline{x} = \{y \in X : yRx\}$. Hence if $x \in A$, for some $A \in X^*$ then $\overline{x} = A$. Now it is easy to see that for

$U \subseteq X^*$, $U \in \mathcal{J}^*$ if and only if $\bigcup_{A \in U} A$ is an open subset of X. Let $(X, \mathcal{J})$ be a topological space and X^* be a family of disjoint nonempty subsets of X such that $X = \bigcup_{A \in X^*} A$. Define $q : X \to X^*$ as $q(x) = A$, where $A \in X^*$ is such that $x \in A$. Then the topology $\mathcal{J}_q$ on X^* is the largest topology on X^* which makes $q : (X, \mathcal{J}) \to (X^*, \mathcal{J}_q)$ a continuous function is called the quotient topology (or identification) topology on X^* induced by q.

Theorem 5. Let $(X^*, \mathcal{J}_q)$ be an identification space (i.e $\mathcal{J}_q$ is the identification topology on X^* with respect to q) defined as above and $(Y, \mathcal{J}')$ be an arbitrary topological space. Then a function $f : (X^*, \mathcal{J}_q) \to (Y, \mathcal{J}')$ is continuous if and only if $f \circ q : (X^*, \mathcal{J}_q) \to (Y, \mathcal{J}')$ is continuous.

Proof: Let $f : (X^*, \mathcal{J}_q) \to (Y, \mathcal{J}')$ be a continuous function. We know that by the definition of identification space, $q : (X, \mathcal{J}) \to (X, \mathcal{J}_q)$ is a continuous function. Hence the composite function $f \circ q : (X^*, \mathcal{J}_q) \to (Y, \mathcal{J}')$ is a continuous function. We will have to prove that $f : (X^*, \mathcal{J}_q) \to (Y, \mathcal{J}')$ is continuous. So start with an open set U in Y. That is we will have to prove that $f^{-1}(U)$ is open in $(X^*, \mathcal{J}_q)$. But the subset $f^{-1}(U)$ is open in the identification space if and only if $q^{-1}(f^{-1}(U))$ is open in $(X, \mathcal{J})$. But $q^{-1}(f^{-1}(U)) = (f \circ q)^{-1}(U)$ an open set in $(X, \mathcal{J})$ (since $f \circ q : (X, \mathcal{J}) \to (Y, \mathcal{J}')$ is a continuous function). This is what we wanted to prove and hence $f : (X, \mathcal{J}_q) \to (Y, \mathcal{J}')$ is a continuous function.

Let $(X, \mathcal{J})$ be a topological space and Y be a nonempty set. Let $f : X \to Y$ be an onto map. Then $X^* = \{f^{-1}(y) : y \in Y\}$ is a family of disjoint subsets of X such that $\bigcup_{y \in Y} f^{-1}(y) = X$. That is X^* is a partition of X. Let $q : X \to X^*$ be the map, known as identification map, defined as above. Let $\mathcal{J}'$ be the largest topology on Y for which $f : (X, \mathcal{J}) \to (Y, \mathcal{J}')$ is continuous. Then it is easy to prove the following:

Theorem 6. Let $(X, \mathcal{J}), (Y, \mathcal{J}')$ be topological spaces and $f : (X, \mathcal{J}) \xrightarrow{onto} (Y, \mathcal{J}')$ be a homeomorphism. Further suppose $(Z, \mathcal{J}_1)$ is any topological space. Then a function $g : (Y, \mathcal{J}') \to (Z, \mathcal{J}_1)$ is continuous if and only if $g \circ f : (X, \mathcal{J}) \to (Z, \mathcal{J}_1)$ is continuous.

Proof: To prove $(Y, \mathcal{J}')$ and $(X^*, \mathcal{J}_q)$ are homeomorphic we will have to define a map say $h : X^* \to Y$ and prove that this map is a homeomorphism.

Let $z \in X^* = \{f^{-1}(y) : y \in Y\}$ be any element. Then $z = f^{-1}(y)$, for some $y \in Y$. So let $h(z)=h(f^{-1}(y))=y$. The defined map $h : X \to Y$ is such that $(h \circ q)(x)=h(q(x))=h(f^{-1}(y))$ (where $y \in Y$ is such that $x \in f^{-1}(y)$)$=y=f(x)$. That is $h \circ q = f$. Let us prove that h is continuous. Let V be an open set in Y.

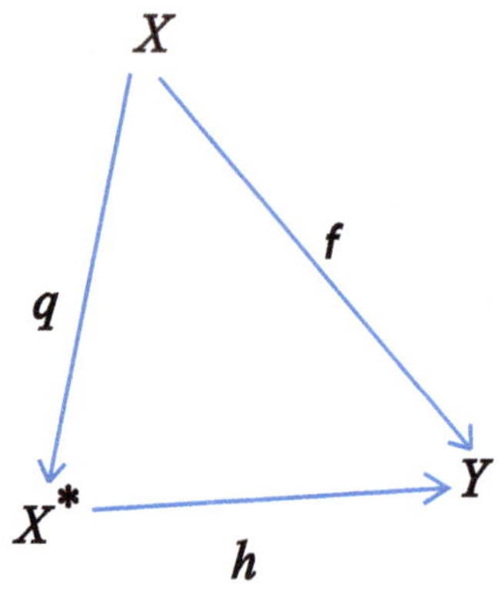

The given topology $\mathcal{J}'$ on Y is the largest topology on Y for which $f:(X, \mathcal{J}) \to (Y, \mathcal{J}')$ is continuous. Hence V is an open set on Y implies $f^{-1}(V)$ is an open set in X and hence $(h \circ q)^{-1}(V) = q^{-1}(h^{-1}(V))$ is an open set in X. Therefore $h^{-1}(V)$ is an open set in X^*. That is V is an open set in Y implies $h^{-1}(V)$ is an open set in X^*. This implies that h is a continuous map.

- The Torus

Let $X = [0, 1] \times [0, 1]$ with the topology $\mathcal{J}$ on X induced by the standard topology on $\mathbb{R}^2$ (that is $\mathcal{J}$ is the topology on X induced by the Euclidean metric). Partition X into the subsets of the type:

- the set $A = \{(0, 0),(0, 1),(1, 0),(1, 1)\}$ consisting of the four corner points,
- all the sets of the form $A_x = \{(x, 0),(x, 1)\}$ for $0 < x < 1$,
- all the sets of the form $A_y = \{(0, y),(1, y)\}$ for $0 < y < 1$,
- all singleton sets of the form $\{(x, y)\}$, $0 < x < 1$, $0 < y < 1$. Then the resulting identification space is the torus.

Exercise: Let $(X, \mathcal{J}),(Y, \mathcal{J}')$ be topological spaces and $f : X \to Y$ be an onto map. If f maps open sets in X to open sets in Y (that is f is an open map) then prove that $\mathcal{J}'$ is the quotient topology on Y induced by f.

3

Connectedness: A Topological Property

When spaces other than disjointed open subsets are connected together, it is said to be a connected topological space. Spaces can be arc-connected, path-connected and locally connected. Topology is best understood in confluence with the major topics listed in the following chapter.

Connected Space

In topology and related branches of mathematics, a connected space is a topological space that cannot be represented as the union of two or more disjoint nonempty open subsets. Connectedness is one of the principal topological properties that are used to distinguish topological spaces.

Connected and Disconnected Subspaces of R^2

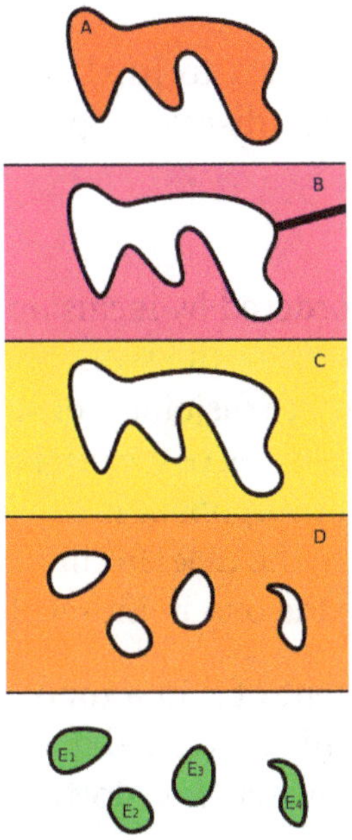

From top to bottom: red space A, pink space B, yellow space C and orange space D are all connected, whereas green space E (made of subsets E1, E2, E3, and E4) is not connected. Furthermore, A and B are also simply connected (genus 0), while C and D are not: C has genus 1 and D has genus 4.

A subset of a topological space X is a connected set if it is a connected space when viewed as a subspace of X.

Formal Definition

A topological space X is said to be disconnected if it is the union of two disjoint non-empty open sets. Otherwise, X is said to be connected. A subset of a topological space is said to be connected if it is connected under its subspace topology. Some authors exclude the empty set (with its unique topology) as a connected space, but this article does not follow that practice.

For a topological space X the following conditions are equivalent:

1. X is connected, that is, it cannot be divided into two disjoint nonempty open sets.
2. X cannot be divided into two disjoint nonempty closed sets.
3. The only subsets of X which are both open and closed (clopen sets) are X and the empty set.
4. The only subsets of X with empty boundary are X and the empty set.
5. X cannot be written as the union of two nonempty separated sets (sets for which each is disjoint from the other's closure).
6. All continuous functions from X to $\{0,1\}$ are constant, where $\{0,1\}$ is the two-point space endowed with the discrete topology.

Connected Components

The maximal connected subsets (ordered by inclusion) of a nonempty topological space are called the connected components of the space. The components of any topological space X form a partition of X: they are disjoint, nonempty, and their union is the whole space. Every component is a closed subset of the original space. It follows that, in the case where their number is finite, each component is also an open subset. However, if their number is infinite, this might not be the case; for instance, the connected components of the set of the rational numbers are the one-point sets (singletons), which are not open.

Let Γ_x be the connected component of x in a topological space X, and Γ'_x be the intersection of all clopen sets containing x (called quasi-component of x.) Then $\Gamma_x \subset \Gamma'_x$ where the equality holds if X is compact Hausdorff or locally connected.

Disconnected Spaces

A space in which all components are one-point sets is called totally disconnected. Related to this property, a space X is called totally separated if, for any two distinct elements

x and y of X, there exist disjoint open sets U containing x and V containing y such that X is the union of U and V. Clearly any totally separated space is totally disconnected, but the converse does not hold. For example take two copies of the rational numbers Q, and identify them at every point except zero. The resulting space, with the quotient topology, is totally disconnected. However, by considering the two copies of zero, one sees that the space is not totally separated. In fact, it is not even Hausdorff, and the condition of being totally separated is strictly stronger than the condition of being Hausdorff.

Examples

- The closed interval [0, 2] in the standard subspace topology is connected; although it can, for example, be written as the union of [0, 1) and [1, 2], the second set is not open in the chosen topology of [0, 2].
- The union of [0, 1) and (1, 2] is disconnected; both of these intervals are open in the standard topological space [0, 1) ∪ (1, 2].
- (0, 1) ∪ {3} is disconnected.
- A convex set is connected; it is actually simply connected.
- A Euclidean plane excluding the origin, (0, 0), is connected, but is not simply connected. The three-dimensional Euclidean space without the origin is connected, and even simply connected. In contrast, the one-dimensional Euclidean space without the origin is not connected.
- A Euclidean plane with a straight line removed is not connected since it consists of two half-planes.
- $\mathbb{R}$, The space of real numbers with the usual topology, is connected.
- If even a single point is removed from $\mathbb{R}$, the remainder is disconnected. However, if even a countable infinity of points are removed from $\mathbb{R}$, where $n \geq 2$, the remainder is connected.
- Any topological vector space over a connected field is connected.
- Every discrete topological space with at least two elements is disconnected, in fact such a space is totally disconnected. The simplest example is the discrete two-point space.
- On the other hand, a finite set might be connected. For example, the spectrum of a discrete valuation ring consists of two points and is connected. It is an example of a Sierpiński space.
- The Cantor set is totally disconnected; since the set contains uncountably many points, it has uncountably many components.

- If a space X is homotopy equivalent to a connected space, then X is itself connected.
- The topologist's sine curve is an example of a set that is connected but is neither path connected nor locally connected.
- The general linear group $\mathrm{GL}(n, R)$ (that is, the group of n-by-n real, invertible matrices) consists of two connected components: the one with matrices of positive determinant and the other of negative determinant. In particular, it is not connected. In contrast, $\mathrm{GL}(n, C)$ is connected. More generally, the set of invertible bounded operators on a (complex) Hilbert space is connected.
- The spectra of commutative local ring and integral domains are connected. More generally, the following are equivalent
    1. The spectrum of a commutative ring R is connected
    2. Every finitely generated projective module over R has constant rank.
    3. R has no idempotent $\neq 0,1$ (i.e., R is not a product of two rings in a nontrivial way).

An example of a space that is not connected is a plane with an infinite line deleted from it. Other examples of disconnected spaces (that is, spaces which are not connected) include the plane with an annulus removed, as well as the union of two disjoint closed disks, where all examples of this paragraph bear the subspace topology induced by two-dimensional Euclidean space.

Path Connectedness

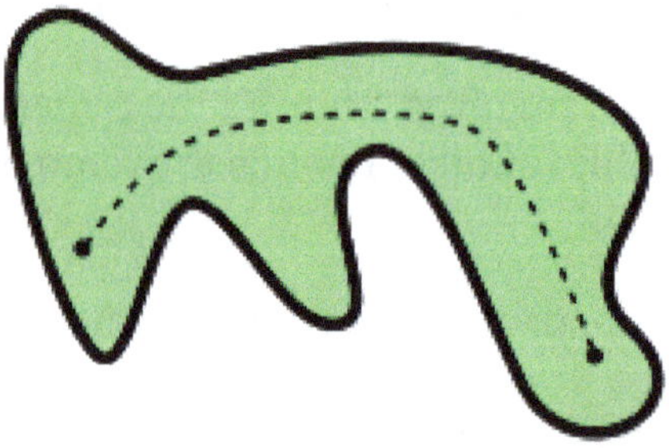

This subspace of R^2 is path-connected, because a path can be drawn between any two points in the space.

A path-connected space is a stronger notion of connectedness, requiring the structure of a path. A path from a point x to a point y in a topological space X is a continuous function f from the unit interval [0,1] to X with f(0) = x and f(1) = y. A path-component of X is an equivalence class of X under the equivalence relation which makes x equivalent to y if there is a path from x to y. The space X is said to be path-connected (or pathwise connected or 0-connected) if there is exactly one path-component, i.e. if there is a path joining any two points in X. Again, many authors exclude the empty space.

Every path-connected space is connected. The converse is not always true: examples of connected spaces that are not path-connected include the extended long line L^* and the *topologist's sine curve.*

Subsets of the real line R are connected if and only if they are path-connected; these subsets are the intervals of R. Also, open subsets of R^n or C^n are connected if and only if they are path-connected. Additionally, connectedness and path-connectedness are the same for finite topological spaces.

Arc Connectedness

A space X is said to be arc-connected or arcwise connected if any two distinct points can be joined by an *arc*, that is a path f which is a homeomorphism between the unit interval [0, 1] and its image $f([0, 1])$. It can be shown any Hausdorff space which is path-connected is also arc-connected. An example of a space which is path-connected but not arc-connected is provided by adding a second copy 0' of 0 to the nonnegative real numbers [0, ∞). One endows this set with a partial order by specifying that $0'<a$ for any positive number a, but leaving 0 and 0' incomparable. One then endows this set with the *order topology*, that is one takes the open intervals $(a, b) = \{x \mid a < x < b\}$ and the half-open intervals $[0, a) = \{x \mid 0 \leq x < a\}$, $[0', a) = \{x \mid 0› \leq x < a\}$ as a base for the topology. The resulting space is a T_1 space but not a Hausdorff space. Clearly 0 and 0' can be connected by a path but not by an arc in this space.

Local Connectedness

A topological space is said to be locally connected at a point x if every neighbourhood of x contains a connected open neighbourhood. It is locally connected if it has a base of connected sets. It can be shown that a space X is locally connected if and only if every component of every open set of X is open. The topologist's sine curve is an example of a connected space that is not locally connected.

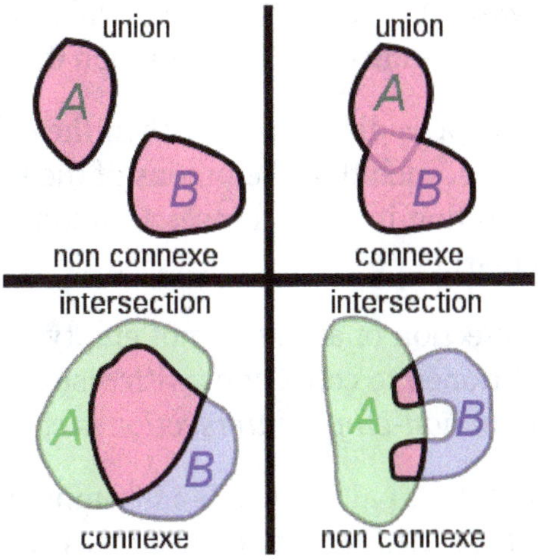

Examples of unions and intersections of connected sets

Similarly, a topological space is said to be locally path-connected if it has a base of path-connected sets. An open subset of a locally path-connected space is connected if and only if it is path-connected. This generalizes the earlier statement about R^n and C^n, each of which is locally path-connected. More generally, any topological manifold is locally path-connected.

Neither local connectedness nor local path connectedness necessarily implies connectedness or path connectedness. For example, the space $(0,1)\cup(2,3)$ is locally connected and locally path connected but neither connected nor path connected.

Set Operations

The intersection of connected sets is not necessarily connected.

Each ellipse is a connected set, but the union is not connected, since it can be partitioned to two disjoint open sets U and V.

The union of connected sets is not necessarily connected. Consider a collection $\{X_i\}$ of connected sets whose union is $X=\cup_i X_i$. If X is disconnected and $X=\cup_i X_i$ is a separation of X (with U,V disjoint and open in X), then each X_i must be entirely contained in either U or V, since otherwise, $X_i\cap U$ and $X_i\cap V$ (which are disjoint and open in X_i) would be a separation of X_i, contradicting the assumption that it is connected.

This means that, if the union X is disconnected, then the collection $\{X_i\}$ can be partitioned to two sub-collections, such that the unions of the sub-collections are disjoint and open in X. This implies that in several cases, a union of connected sets *is* necessarily connected. In particular:

1. If the common intersection of all sets is not empty ($\cap X_i \neq \varnothing$), then obviously they cannot be partitioned to collections with disjoint unions. Hence *the union of connected sets with non-empty intersection is connected.*

2. If the intersection of each pair of sets is not empty ($\forall i,j: X_i\cap X_j \neq \varnothing$) then again they cannot be partitioned to collections with disjoint unions, so their union must be connected.

3. If the sets can be ordered as a "linked chain", i.e. indexed by integer indices and $\forall i: X_i \cap X_{i+1} \neq \varnothing$, then again their union must be connected.

4. If the sets are pairwise-disjoint and the quotient space $X / \{X_i\}$ is connected, then X must be connected. Otherwise, if $U \cup V$ is a separation of X then $q(U) \cup q(V)$ is a separation of the quotient space (since $q(U), q(V)$ are disjoint and open in the quotient space).

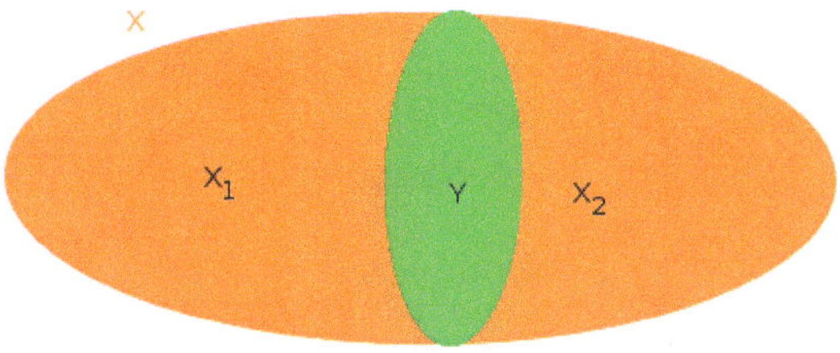

Two connected sets whose difference is not connected

The set difference of connected sets is not necessarily connected. However, if $X \supseteq Y$ and their difference $X \setminus Y$ is disconnected (and thus can be written as a union of two open sets X_1 and X_2), then the union of Y with each such component is connected (i.e. $Y \cup X_i$ is connected for all i). Proof: By contradiction, suppose $Y \cup X_1$ is not connected. So it can be written as the union of two disjoint open sets, e.g. $Y \cup X_1 = Z_1 \cup Z_2$. Because Y is connected, it must be entirely contained in one of these components, say Z_1, and thus Z_2 is contained in X_1. Now we know that:

$$X = (Y \cup X_1) \cup X_2 = (Z_1 \cup Z_2) \cup X_2 = (Z_1 \cup X_2) \cup (Z_2 \cap X_1)$$

The two sets in the last union are disjoint and open in X, so there is a separation of X, contradicting the fact that X is connected.

Theorems

- Main theorem of connectedness: Let X and Y be topological spaces and let $f: X \to Y$ be a continuous function. If X is (path-)connected then the image $f(X)$ is (path-)connected. This result can be considered a generalization of the intermediate value theorem.
- Every path-connected space is connected.
- Every locally path-connected space is locally connected.
- A locally path-connected space is path-connected if and only if it is connected.
- The closure of a connected subset is connected.
- The connected components are always closed (but in general not open)
- The connected components of a locally connected space are also open.

- The connected components of a space are disjoint unions of the path-connected components (which in general are neither open nor closed).
- Every quotient of a connected (resp. locally connected, path-connected, locally path-connected) space is connected (resp. locally connected, path-connected, locally path-connected).
- Every product of a family of connected (resp. path-connected) spaces is connected (resp. path-connected).
- Every open subset of a locally connected (resp. locally path-connected) space is locally connected (resp. locally path-connected).
- Every manifold is locally path-connected.

Graphs

Graphs have path connected subsets, namely those subsets for which every pair of points has a path of edges joining them. But it is not always possible to find a topology on the set of points which induces the same connected sets. The 5-cycle graph (and any n-cycle with $n>3$ odd) is one such example.

As a consequence, a notion of connectedness can be formulated independently of the topology on a space. To wit, there is a category of connective spaces consisting of sets with collections of connected subsets satisfying connectivity axioms; their morphisms are those functions which map connected sets to connected sets (Muscat & Buhagiar 2006). Topological spaces and graphs are special cases of connective spaces; indeed, the finite connective spaces are precisely the finite graphs.

However, every graph can be canonically made into a topological space, by treating vertices as points and edges as copies of the unit interval. Then one can show that the graph is connected (in the graph theoretical sense) if and only if it is connected as a topological space.

Stronger Forms of Connectedness

There are stronger forms of connectedness for topological spaces, for instance:

- If there exist no two disjoint non-empty open sets in a topological space, X, X must be connected, and thus hyperconnected spaces are also connected.
- Since a simply connected space is, by definition, also required to be path connected, any simply connected space is also connected. Note however, that if the "path connectedness" requirement is dropped from the definition of simple connectivity, a simply connected space does not need to be connected.

- Yet stronger versions of connectivity include the notion of a contractible space. Every contractible space is path connected and thus also connected.

In general, note that any path connected space must be connected but there exist connected spaces that are not path connected. The deleted comb space furnishes such an example, as does the above-mentioned topologist's sine curve.

Connected Spaces

Definition: A topological space $(X, \mathcal{J})$ is said to be a disconnected topological space if there exist nonempty open sets A and B of X such that (i) $A \cap B = \phi$, (ii) X = A ∪ B.

In such a case $B = A^c$ and $A = B^c$ and hence A and B are closed sets. Also X contains a nonempty proper subset A (that is $A \neq \phi$, X which is both open and closed in X.

A topological space $(X, \mathcal{J})$ is said to be connected if there cannot exist nonempty closed (open) subsets A and B of X such that (i) $A \cap B = \phi$, (ii) X = A∪B. Equivalently, $(X, \mathcal{J})$ is connected if and only if ϕ and X are the only subsets of X which are both open and closed in X.

Examples. (i) Let X be a set containing at least two elements and A ⊆ X, $A \neq \phi$, X. Then $\mathcal{J} = \{\phi, X, A, A^c\}$ is a topological space. Here A is such that $A \neq \phi$, $A \neq X$ and A is both open and closed. Hence $(X, \mathcal{J})$ is not a connected topological space.

(ii) Let X be a nonempty set and $\mathcal{J}_f = \{A \subseteq X, X \setminus A = A^c$ finite or $A^c = X\}$ then $(X, \mathcal{J}_f)$ is a topology on X, known as cofinite topology on X.

Note. If X is a finite set containing at least two elements then $\mathcal{J}_f$ is the discrete topology on X. In this case $(X, \mathcal{J}_f)$ is not a connected topological space.

What will happen when X is an infinite set and $\mathcal{J}_f$ is the cofinite topology on X. Is $(X, \mathcal{J}_f)$ not connected? In other words can we find a subset A of X such that $A \neq \phi$ and $A \neq X$ but A is both open and closed? $A \neq X$, A is closed implies A is a finite set. Also $A \neq \phi$ implies such a nonempty finite set cannot be an open set. Therefore ϕ and X are the only sets which are both open and closed. This implies $(X, \mathcal{J}_f)$ is a connected topological space whenever X is an infinite set.

Connected Subsets of the Real Line

Keeping our intuition alive, let us prove that intervals are connected subsets of $\mathbb{R}$ and they are the only connected subsets of $\mathbb{R}$. Recall that a subset J of $\mathbb{R}$ is an interval if and only if whenever $a, b \in J$ and $a < c < b$, we have $c \in J$. Note that the null set ϕ and singleton sets are also intervals. For example for $x \in \mathbb{R}$, $\phi = (x, x) = (1, 1)$ and {x} = [x, x], the singleton set containing x.

We say that a subset Y of a topological space X is connected if $(Y, \mathcal{J}_Y)$ is connected.

First let us prove that if $A \subseteq \mathbb{R}$ is not an interval then A is not connected. Here $\mathcal{J}_A = \{A \cap U : U$ is an open set in $\mathbb{R}\}$. So, we will have to prove that the topological space $(A, \mathcal{J}_A)$ is not connected. The given set A is not an interval implies there exist $x, y \in A$ and $z \in \mathbb{R}$ such that $x < z < y$ and $z \notin A$. We know that $(-\infty, z), (z, \infty)$ are open sets in $\mathbb{R}$. This implies that $(-\infty, z) \cap A$ and $(z, \infty) \cap A$ are open sets in $(A, \mathcal{J}_A)$. Also $x \in (-\infty, z) \cap A = C$ and $y \in (z, \infty) \cap A = D$ and $A = C \cup D$. That is C, D are nonempty open sets in $(A, \mathcal{J}_A)$ such that $C \cap D = \phi$ and $A = C \cup D$. Hence the topological space $(A, \mathcal{J}_A)$ cannot be connected. That is the given subset A of $\mathbb{R}$ (which is not an interval) is not connected.

Now let us prove:

Theorem 1. Every interval in $\mathbb{R}$ is connected

Proof: Let J be an interval and let us assume that J contains at least two elements. For if $J = \phi$ or a singleton set then the null set ϕ and the whole space J are the only sets which are both open and closed in J and hence J is connected. Now let us suppose that there exist nonempty closed sets A, B in J (that is A, B $\subseteq$ J and A, B are closed sets in $(J, \mathcal{J}_J)$, where $\mathcal{J}_J = \{U \cap J : U$ is open in $\mathbb{R}\}$) such that $J = A \cup B$. Fix $a \in A$, $b \in B$ and without loss of generality let us say $a < b$. Note that $a \in A \subseteq J$, $b \in B \subseteq J$.

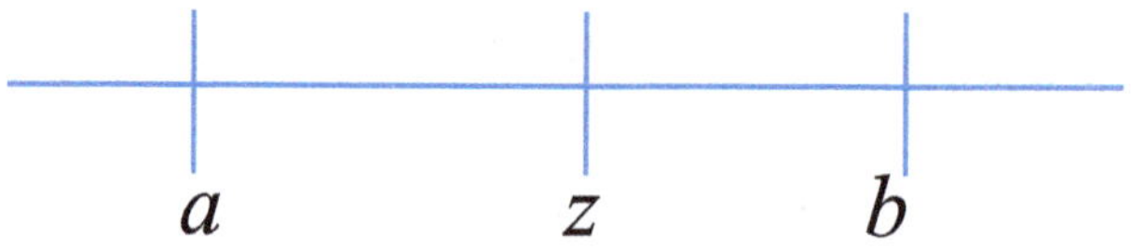

Since J is an interval $[a,b] \subseteq J$. Let $y=\sup(A \cap [a,b])$. Now let us prove that $y \in A \cap B$. First let us prove that $y \in A$. Let U be an open set in J containing y ($y \in [a, b] \subseteq J$). Then there exists an open set V in $\mathbb{R}$ such that $V \cap J = U$. Hence there exists $\epsilon > 0$ such that $(y-\epsilon, y+\epsilon) \cap J \subseteq V \cap J = U$. (If $y = a$, we are through, otherwise take $0 < \epsilon < y - a$ and $\epsilon < b - y$ when $y \neq b$.) Now $y-\epsilon$ is not an upper bound for $[a, b] \cap A$ implies there exists $x_0 \in [a, b] \cap A$ such that $y - \epsilon < x_0$. This implies $x_0 \in (y - \epsilon, y + \epsilon) \cap A \subseteq (y-\epsilon, y+\epsilon) \cap J \subseteq U$. Hence $x_0 \in U \cap A$. That is, whenever U is an open set in J containing y, then $U \cap A \neq \phi$. This proves that $y \in \overline{A}_J = A$, $\overline{A}_J$ is the closure of A in J. Also if $y = b$, then $y \in B$. Suppose $y < b$, (y is the supremum of $C = A \cap [a, b]$ and b is an upper bound of C then take $y_0 \in (y, y+\epsilon)$. Now $y_0 \in U \cap B$. Hence $y \in \overline{B}_J = B$. Hence we have $y \in A \cap B$. Therefore there cannot exist nonempty closed subsets of A, B in the subspace $(J, \mathcal{J}_J)$ such that $A \cap B = \phi$, $A \cup B = J$. This proves that J is connected.

Some Properties of Connected Spaces

Now let us prove that continuous image of a connected topological space is connected.

Theorem 2. Let $(X, \mathcal{J})$ be a connected topological space and $(Y, \mathcal{J}')$ be any topological space. Suppose $f:(X, \mathcal{J}) \to (Y, \mathcal{J}')$ is a surjective continuous map then the image f(X) = Y is a connected topological space.

Proof: (By contradiction). Let A, B be nonempty open subsets of Y such that $A \cap B = \phi,\ A \cup B = Y$. Now f is a continuous map, and A, B are open sets in Y implies that $f^{-1}(A)$, $f^{-1}(B)$ are open sets in X. Also

$$f^{-1}(A) \cap f^{-1}(B) = f^{-1}(A \cap B) = \phi \tag{1}$$

And

$$\begin{aligned} X &= f^{-1}(Y) = f^{-1}(A \cup B) \\ &= f^{-1}(A) \cup f^{-1}(B). \end{aligned} \tag{2}$$

Since $A,\ B \neq \phi$, let $y \in A$ and $y' \in B$, and f is a surjective map implies that there exist $x,\ x' \in X$ such that f(x) = y and $f(x') = y'$ implies $f(x) \in A$ and $f(x') \in B$ implies $x \in f^{-1}(A)$ and $x' \in f^{-1}(B)$. Hence $f^{-1}(A)$, $f^{-1}(B)$ are nonempty open subsets of the connected topological space $(X,\ \mathcal{J})$ satisfying Eqs. (1) and (2). This means $(X,\ \mathcal{J})$ cannot be a connected topological space. That is we have proved: f(X) = Y is not connected implies X is also not connected and this gives a contradiction. Hence our assumption that $(Y,\ \mathcal{J}')$ is not connected is not valid. Therefore $(Y,\ \mathcal{J}')$ is a connected topological space.

It is to be noted that if X is a connected topological space and $f : X \to Y$ is a continuous function, where Y is any topological space, then the image f(X) is also a connected topological space. Here we will have to consider f(X) as a subspace of the given topological space Y. Also if $f : X \to Y$ is continuous then $f : X \to f(X)$ is also continuous. Hence this result will follow from the previous result.

Definition: A subspace Y of a topological space $(X,\ \mathcal{J})$ is said to have a separation if and only if there exist nonempty subsets A, B of X such that (i) Y = A ∪ B, (ii) $\overline{A} \cap B = \phi = A \cap \overline{B}$. Here $\overline{A}$ is the closure of A in X.

Example: Let X = $\mathbb{R}$ and Y = [0, 2) ∩ (2, 5). Then Y has a separation. Take A = [0, 2), B = (2, 5), then (i) Y = A ∪ B is satisfied. Now (ii) $\overline{A} \cap B = [0,\ 2] \cap (2,5) = \phi$, and $A \cap \overline{B} = [0,\ 2) \cap (2,5) = \phi$. Hence from (i) and (ii) we see that Y has a separation.

Theorem 3. A subspace Y of a topological space $(X,\ \mathcal{J})$ is connected if and only if there does not exist any separation for Y.

Proof: First let us assume that the subspace Y is connected. This means the subspace $(Y,\ \mathcal{J}_Y)$ is a connected topological space. So, we will have to prove that Y does not admit any separation. Suppose Y has a separation. Hence there exist nonempty subsets A, B of X such that $Y = A \cup B,\ \overline{A} \cap B = \phi = A \cap \overline{B}$. Now $\overline{A}_Y = \overline{A} \cap Y = \overline{A} \cap (A \cup B) - (\overline{A} \cap A) \cup (\overline{A} \cap B) = A \cup \phi = A$. This implies that A is a closed subset of $(Y,\ \mathcal{J}_Y)$. Similarly $\overline{B}_Y = \overline{B} \cap Y = (\overline{B} \cap (A \cup B)) = (\overline{B} \cap A) \cup (\overline{B} \cap B) = B$.

This implies that B is a closed set in (Y, J_Y). Also $B_Y^c = Y \setminus B = Y \cap B^c = (A \cup B) \cap B^c$ $= (A \cap B^c) \cup (B^c \cap B) = A \cap B^c = A$ (since $A \cap B = \phi$). This means complement of B with respect to Y is A. Hence B is such that $B \neq \phi, B \neq Y$ (since Y = A ∩ B and $A \neq \phi$) and B is both open and closed in Y. This implies that the subspace (Y, J_Y) is a disconnected space. This is a contradiction and we arrived at this contradiction by assuming that Y has a separation. Hence there does not exist any separation for Y.

Conversely, now assume that there does not exist any separation for Y. Now suppose that the subspace (Y, J_Y) is a disconnected space. Then there exist nonempty closed subsets A, B in (Y, J_Y) such that Y = A ∪ B and $A \cap B = \phi$. Now $\bar{A} \cap B = \bar{A} \cap (Y \cap B) = (\bar{A} \cap Y) \cap B = \bar{A}_Y \cap B = A \cap B = \phi$. (A is closed in Y implies $\bar{A}_Y = A$) Similarly, $A \cap \bar{B} = A \cap Y \cap \bar{B} = A \cap \bar{B}_Y = A \cap B = \phi$. (B is closed in Y implies $\bar{B}_Y = B$.) We have nonempty subsets A, B in X such that (i) Y = A ∪ B, (ii) $\bar{A} \cap B = \phi = A \cap \bar{B}$. This means Y has a separation and this gives a contradiction. We arrived at this contradiction by assuming that the subspace (Y, J_Y) is a disconnected space. Hence the assumption is wrong. That means (Y, J_Y) is connected.

Theorem 4. Let (X, J) be a disconnected topological space and A be a subset of X such that (i) $A \neq \phi, X$ (ii) A is both open and closed in X. Suppose Y is a nonempty connected subspace of X. Then either Y ⊆ A or Y ⊆ A^c.

Proof: X = A ∪ B, where B = A^c implies Y = X ∩ Y = (A ∪ B) ∩ Y = (A ∩ Y) ∪ (B ∩ Y). Also $(A \cap Y) \cap \overline{(B \cap Y)} \subseteq A \cap \bar{B} = A \cap B = \phi$ (since $B = A^c$ is a closed set in X) implies $(A \cap Y) \cap \overline{(B \cap Y)} = \phi$. Similarly, $\overline{(A \cap Y)} \cap (B \cap Y) \subseteq \bar{A} \cap B = A \cap B = \phi$ implies $\overline{(A \cap Y)} \cap (B \cap Y) = \phi$. It is given that Y is a connected subspace of X. Hence it cannot happen that $A \cap Y \neq \phi$ and $B \cap Y \neq \phi$. Since Y is a connected subspace of X, Y cannot admit any separation. Hence $A \cap Y = \phi$ or $B \cap Y = \phi$ implies $Y \subseteq A^c$ or $Y \subseteq B^c = A$.

Alternate Proof: Now $A \cap Y, B \cap Y$ (where B = A^C) are open sets in the subspace Y such that (i) $Y = (A \cap Y) \cup (B \cap Y)$ and (ii) $(A \cap Y) \cap (B \cap Y) \subseteq A \cap B = \phi$ implies $A \cap Y = \phi$ or $B \cap Y = \phi$ implies $Y \subseteq A^C$ or $Y \subseteq C$. Let us prove the following:

Theorem 5. Let (X, J) be a topological space and Y_1, Y_2 be connected subspaces of X. Further suppose $Y_1 \cap Y_2 \neq \phi$. Then $Y_1 \cup Y_2$ is a connected subspace of X.

Proof: We will have to prove that the subspace $Y_1 \cup Y_2$ cannot admit any separation. Suppose A, B are subsets of X such that (i) $Y_1 \cup Y_2 = A \cup B$, (ii) $\bar{A} \cap B = \phi = A \cap \bar{B}$ (here $\bar{A}$ is the closure of A in X).

From (i)

$$Y_1 = (Y_1 \cap A) \cup (Y_1 \cap B), \tag{3}$$

and from (ii)

$$\overline{(Y_1 \cap A)} \cap (Y_1 \cap B) = \phi \text{ and } (Y_1 \cap A) \cap \overline{(Y \cap B)} = \phi. \tag{4}$$

It is given that Y_1 is a connected subspace. Hence Y_1 cannot admit a separation. Therefore from Eqs. (3) and (4) we conclude that $Y_1 \cap A = \phi$ or $Y_1 \cap B = \phi$. Let us say that $Y_1 \cap A = \phi$. This implies that

$$Y_1 \subseteq A^c. \tag{5}$$

Similarly, using the fact that Y_2 is a connected subspace we conclude that $Y_2 \cap A = \phi$ or $Y_2 \cap B = \phi$. If $Y_2 \cap B = \phi$ then

$$Y_2 \subseteq B^c. \tag{6}$$

From Eqs. (5) and (6) we have

$$Y_1 \cap Y_2 \subseteq A^c \cap B^c = (A \cup B)^c. \tag{7}$$

Also $Y_1 \cap Y_2 \neq \phi$. Hence there exists an element say $x_0 \in Y_1 \cap Y_2$. Then from Eq. (7), $x_0 \notin A \cup B$, a contradiction to $Y_1 \cup Y_2 = A \cup B$. Therefore both $Y_1 \subseteq A^c$ and $Y_2 \subseteq A^c$ implies $Y_1 \cup Y_2 \subseteq A^c$ implies $A = \phi$. (If there exists $x \in A \subseteq Y_1 \cup Y_2$ then $x \in Y_1 \cup Y_2 \subseteq A^c$.) What we have proved is the following: If (i) and (ii) happens then that leads to $A = \phi$ or $B = \phi$. ($Y_1 \cap A = \phi$ or $Y_1 \cap B = \phi$ implies $Y_1 \subseteq A^c$ or $Y_1 \subseteq B^c$. So if we assume $Y_1 \subseteq B^c$ then we would have arrived at $B = \phi$.) That is we have proved that $Y_1 \cup Y_2$ cannot admit any separation and this implies $Y_1 \cup Y_2$ is a connected subspace of X.

Remark : If X is a topological space and $Y_1, Y_2, \ldots, Y_n$ for some $n \in \mathbb{N}$ is a finite collection of connected topological spaces such that $\bigcap_{i=1}^{n} Y_i \neq \phi$ then by using induction we can prove that $Y_1 \cup Y_2 \cup \cdots \cup Y_n$ is a connected subset of X.

In fact using exactly the same idea of proving that the subspace $Y_1 \cup Y_2$ is connected whenever Y_1, Y_2 are connected subspaces with the added condition that $Y_1 \cap Y_2 \neq \phi$, we prove the following theorem.

Theorem 6. If $\{Y_\alpha\}_{\alpha \in J}$ is a collection of connected subspaces of a topological space X and further there exists an $x_0 \in X$ such that $x_0 \in Y_\alpha$, for each $\alpha \in J$, then $\bigcup_{\alpha \in J} Y_\alpha$ is a connected subspace of X.

Proof: Here again we will prove that $\bigcup_{\alpha\in J} Y_\alpha$ cannot admit any separation. Suppose A, B are subsets of X such that (i) $\bigcup_{\alpha\in J} Y_\alpha = A\cup B$, (ii) $\overline{A}\cap B = A\cap\overline{B} = \phi$. Then our aim is to prove $A=\phi$ or $B=\phi$. For each fixed $\alpha_0 \in J$, $Y_{\alpha_0} = \left(Y_{\alpha_0}\cap A\right)\cup\left(Y_{\alpha_0}\cap B\right)$ and $\overline{\left(Y_{\alpha_0}\cap A\right)}\cap\left(Y_{\alpha_0}\cap B\right) = \phi = \left(Y_{\alpha_0}\cap A\right)\cap\overline{\left(Y_{\alpha_0}\cap B\right)}$. Hence Y_{α_0} is a connected subspace of X implies $Y_{\alpha_0}\cap A=\phi$ or $Y_{\alpha_0}\cap B=\phi$. Let us say $Y_{\alpha_0}\cap A=\phi$. This will imply that $Y_{\alpha_0}\subseteq A^c$. Note that $\alpha_o \in J$ is an arbitrary element. Hence for each $\alpha \in J$, $Y_\alpha \subseteq A^c$ or $Y_\alpha \subseteq B^c$. But if $Y_{\alpha_1}\subseteq A^c$ for some $\alpha_1 \in J$ and $Y_{\alpha_2}\subseteq B^c$ for some $\alpha_2 \in J$ then

$$x_0 \in Y_{\alpha_1}\cap Y_{\alpha_2} \subseteq A^c\cap B^c = (A\cup B)^c = \left(\bigcup_{\alpha\in J} Y_\alpha\right)^c$$

. This means $x_0 \notin Y_\alpha$ for some $\alpha \in J$ and that gives a contradiction. Therefore $Y_\alpha \subseteq A^c$ for all $\alpha \in J$ or $Y_\alpha \subseteq B^c$ for all $\alpha \in J$ implies $\bigcup_{\alpha\in J} Y_\alpha \subseteq A^c$ or $\bigcup_{\alpha\in J} Y_\alpha \subseteq B^c$ implies $A=\phi$ or $B=\phi$. This means $\bigcup_{\alpha\in J} Y_\alpha$ cannot admit any separation and that is what we wanted to prove.

Theorem 7. Let X be a topological space and $Y_1, Y_2, \ldots, Y_n, \ldots$ be a collection of connected topological spaces. Further suppose $Y_k \cap Y_{k+1} \neq \phi$ for all $k\in\mathbb{N}$. Then $\bigcup_{k=1}^{\infty} Y_k$ is also a connected space.

Proof: Now Y_1, Y_2 are connected subspaces of X and $Y_1\cap Y_2 \neq \phi$ implies $E_2 = Y_1 \cup Y_2$ is a connected subspace. Now E_2 is a connected subspace of X and Y_3 is a connected subspace of X. Further $Y_2 \cap Y_3 \subseteq (Y_1 \cup Y_2) \cap Y_3 = E_2 \cap Y_3$. Hence $Y_2\cap Y_3\neq\phi$ implies $E_2\cap Y_3 \neq\phi$ implies $E_3 = E_2 \cup Y_3 = Y_1 \cup Y_2 \cup Y_3$ is a connected subspace of X. Now use induction to prove that, for each $k\in\mathbb{N}$, $E_k = Y_1 \cup Y_2 \cup \cdots \cup Y_k$ is a connected subspace of X. We have a collection $\{E_k\}_{k=1}^{\infty}$ of connected subspaces of X such that $Y_1 = E_1 \subseteq \bigcap_{k=1}^{\infty} E_k$. Also $\bigcap_{k=1}^{\infty} E_k \neq\phi$. Since $Y_1\neq\phi$. Hence $\bigcup_{k=1}^{\infty} E_k = \bigcup_{k=1}^{\infty} Y_k$ is a connected subspace.

Now let us prove that if Y is a nonempty connected subspace of a topological space X and then it remains connected after adding some (or all) of its limits points to E.

Theorem 8. Let Y be a nonempty connected subspace of a topological space X and E be a subset of X such that $Y\subseteq E\subseteq\overline{Y}$. Then E is also a connected subspace of X.

Proof: Let A, B be subsets of X such that (i) E=A ∪ B (ii) $\overline{A}\cap B = A\cap\overline{B}=\phi$ then Y=Y ∩ E = (A ∩ Y) ∪ (B ∩ Y). Also $\overline{(A\cap Y)}\cap(B\cap Y) = \phi = (A\cap Y)\cap\overline{(B\cap Y)}$. Hence Y is a connected subspace of X implies $A\cap Y = \phi$ or $B\cap Y=\phi$. Let us say $A\cap Y=\phi$. This implies $Y = B\cap Y \Rightarrow Y\subseteq B \Rightarrow \overline{Y}\subseteq\overline{B}$. Since, $E\subseteq\overline{Y}$, $E = A\cup B$ and $A\cap\overline{B} = \phi$ we get $A\subseteq E\subseteq\overline{B}$ and $A = A\cap\overline{B}=\phi$. That is we have proved that for subsets A, B of X, $E = A\cup B$, $\overline{A}\cap B = A\cap\overline{B}=\phi$. This implies $A = \phi$. We arrived at this conclusion by assuming that $A\cap Y=\phi$. If we had assumed $B\cap Y=\phi$, then we would have proved $B = \phi$. Therefore E does not admit any separation and hence E is a connected subspace of X.

Theorem 9. If $(X_1, J_1), (X_2, J_2)$ are connected topological spaces, then the product space $X_1 \times X_2$ is also a connected space.

Proof: Fix $a_1 \in X_1$, $a_2 \in X_2$. (Note. $X_1 = \phi$ or $X_2 = \phi \Rightarrow X_1 \times X_2 = \phi$ and in this case $X_1 \times X_2$ is a connected space.) For each $x_1 \in X_1$, $f_{x_1} : X_2 \to X_1 \times X_2$ defined as $f_{x_1}(x_2) = (x_1, x_2)$ is a continuous function. Also we know that continuous image of a connected space is connected. In this case, the continuous image is $f_{x_1}(X_2) = x_1 \times X_2$. That is $x_1 \times X_2 = \{(x_1, x_2) : x_2 \in X_2\}$ is a connected subspace of the product space $X_1 \times X_2$ (refer the vertical line passing through x_1). Similarly, $X_1 \times a_2$ is a connected subspace of the product space. Also $(x_1, a_2) \in (x_1 \times X_2) \cap (X_1 \times a_2)$. Hence $(x_1 \times X_2) \cup (X_1 \times a_2)$ is a connected subspace of the product space. Let $T_{x_1} = (x_1 \times X_2) \cup (X_1 \times a_2)$. Also note that $(a_1, a_2) \in T_{x_1}$, for each $x_1 \in X_1$. That is $\{T_{x_1}\}$ is a collection of connected subspaces of the product space $X_1 \times X_2$. Further $(a_1, a_2) \in T_{x_1}$ for all $x_1 \in X_1$. Hence $\bigcup_{x_1 \in X_1} T_{x_1} = X_1 \times X_2$ is a connected space.

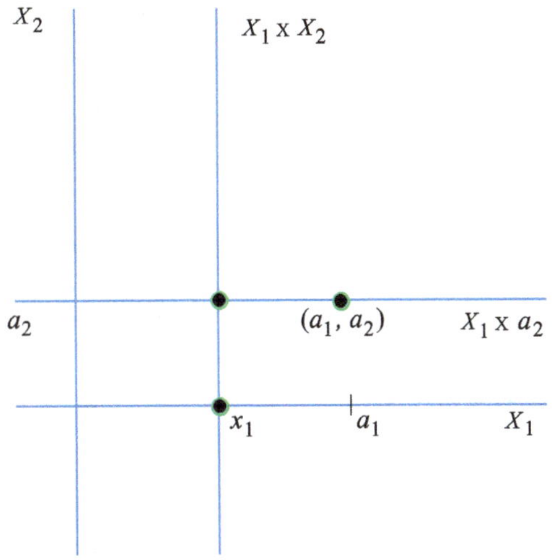

Now we use mathematical induction to prove: if $(X_1, J_1), (X_2, J_2), \ldots, (X_n, J_n)$ are a finite collection of connected topological spaces then the product space $X_1 \times X_2 \times \cdots \times X_n$ is also a connected space. But it is to be noted that we cannot use (say why?) mathematical induction to prove: If (X_n, J_n), $n \in \mathbb{N}$ is a collection of connected topological spaces then the product space $X_1 \times X_2 \times \cdots = \prod_{n=1}^{\infty} X_n$ is also a connected space.

However, we prove the following theorem when we have a collection (X_α, J_α), $\alpha \in J$ (where J is a nonempty index set) of connected topological spaces.

Theorem 10. Let (X_α, J_α), $\alpha \in J$ be a collection of connected topological spaces. Then the product space $X = \prod_{\alpha \in J} X_\alpha$ is also a connected space.

Proof: When J is a finite set we have already proved this result. So let us assume that J is an infinite set. Also let us assume that each $X_\alpha \neq \phi$. From each X_α, fix an element say $a_\alpha \in X_\alpha$. That is we have a function $f : J \to \bigcup_{\alpha \in J} X_\alpha$ such that $f(\alpha) = a_\alpha \in X_\alpha$ for all $\alpha \in J$. That is $f \in \prod_{\alpha \in J} X_\alpha$. We normally write $f = (a_\alpha)_{\alpha \in J}$ and just say that $(a_\alpha)_{\alpha \in J} \in \prod_{\alpha \in J} X_\alpha$. If $\alpha_1, \alpha_2, \ldots, \alpha_k \in J$ then $X(\alpha_1, \alpha_2, \ldots, \alpha_k) = \{x = (x_\alpha)_{\alpha \in J} : x_\alpha = a_\alpha$, when $\alpha \neq \alpha_1, \alpha_2, \ldots, \alpha_k\}$. If our $J = \mathbb{N}$ and $\alpha_j = j$, $j = 1, 2, \ldots, k$ then $X(1, 2, \ldots, k) = \{x_1, x_2, \ldots, x_k, a_{k+1}, a_{k+2}, \ldots\} = X_1 \times X_2 \times \cdots \times X_k \times a_{k+1} \times a_{k+2} \times \cdots$ Note that $g(x_1, x_2, \ldots, x_k) = (x_1, x_2, \ldots, x_k, a_{k+1}, a_{k+2} \ldots)$ is a continuous function from the connected space $X_1 \times X_2 \times \cdots \times X_k \to \prod_{n=1}^{\infty} X_n$ and its image is $X_1 \times X_2 \times \cdots \times X_k \times a_{k+1} \times a_{k+2} \times \cdots$. Hence $X(1, 2, \ldots, k) = X_1 \times X_2 \times \cdots \times X_k \times a_{k+1} \times a_{k+2} \times \cdots$ is a connected subspace of $\prod_{n \in \mathbb{N}} X_n$) and $X(\alpha_1, \alpha_2, \ldots, \alpha_k)$ is a connected subspace of the product space. Also our fixed $a = (a_\alpha)_{\alpha \in J} \in X(\alpha_1, \alpha_2, \ldots, \alpha_k)$. Therefore $Y = \bigcup_{\{\alpha_1, \alpha_2, \ldots, \alpha_k\} \subseteq J} X(\alpha_1, \alpha_2, \ldots, \alpha_k)$ (that is $x \in Y$ if and only if there exists $k \in \mathbb{N}$ and $\alpha_1, \alpha_2 \ldots, \alpha_k \in J$ such that $x \in X(\alpha_1, \alpha_2, \ldots, \alpha_k)$) is a connected subspace of the product space.

Again Y is a connected subspace of the product space implies $\overline{Y}$ is also a connected space. Now let us prove that $\overline{Y} = \prod_{\alpha \in J} X_\alpha = X$. So take an element say $x = (x_\alpha)_{\alpha \in J} \in X$ our aim is to prove that $x \in \overline{Y}$. Start with a basic open set say $U = \prod_{\alpha \in J} U_\alpha$ containing the point $x = (x_\alpha)_{\alpha \in J}$. U is a basic open set in the product space implies there exists $k \in \mathbb{N}$ and $\alpha_1, \alpha_2, \ldots, \alpha_k \in J$ such that $U_\alpha = X_\alpha$ for $\alpha \in J$ and $\alpha \neq \alpha_1, \alpha_2, \ldots, \alpha_k$ and $U_{\alpha_j} \in J_{\alpha_j}$ for all $j = 1, 2, \ldots, k$. Let $y = (y_\alpha)_{\alpha \in J}$ be such that $y_{\alpha_j} = x_{\alpha_j}$ for all $j = 1, 2, \ldots, k$ and $y_\alpha = x_\alpha$ when $\alpha \neq \alpha_1, \alpha_2, \ldots, \alpha_k$. Then $y \in U \cap X(\alpha_1, \alpha_2, \ldots, \alpha_k) \subseteq U \cap Y$ and hence $U \cap Y \neq \phi$ for every basic open set U containing x. This implies $x \in \overline{Y}$. Hence $X \subseteq \overline{Y}$. This implies $X = \overline{Y}$. Now Y is a connected subspace of X implies Y is also a connected subspace of X and hence $X = \overline{Y}$ is a connected space.

Connected Components

Now let us define a relation say R on a given topological space $(X, \mathcal{J})$. We know that a relation on X is a subset of $X \times X$. Here $R = \{(x, y) \in X \times X : x, y \in E_{xy}$ for a connected subset E_{xy} of X$\}$. If $(x, y) \in R$ then we write xRy (read as x is related to y). So we say that xRy if and only if there is a connected set containing x, y. Now it is easy to see that this relation is an equivalence relation. For each $x \in X$, singleton set $\{x\}$ is a connected subset of X. That is $E_{xx} = \{x\}$ is a connected set containing x, x. Hence xRx for all $x \in X$. Therefore R is reflexive.

Now suppose for $x, y \in X$, xRy this implies there is a connected subset say E_{xy} of X such that x, y ∈ E_{xy}. But x, y ∈ E_{xy} implies y, x ∈ E_{xy} this implies yRx. That is for x, y ∈ X, xRy ⇒ yRx and hence R is symmetric.

Now suppose for x, y, z ∈ X xRy and yRz. Now xRy implies there exists a connected subset say E_{xy} of X such that x, y ∈ E_{xy}. Similarly, yRz implies there exists a connected subset say E_{yz} of X such that y, z ∈ E_{yz}. Now E_{xy}, E_{yz} are connected subsets of X such that y ∈ $E_{xy} \cap E_{yz}$. This gives us $E_{xy} \cup E_{yz}$ is a connected set and further y, z ∈ $E_{xy} \cup E_{yz}$. Hence there exists a connected set containing y, z. This implies yRz. That is we have proved that xRy and yRz implies xRz that is R is transitive.

Thus R is an equivalence relation on X. Hence this relation will partition the set X into disjoint equivalence classes.

It is to be noted that for each $x \in X$ there is exactly one equivalence class containing x. Let us denote the equivalence class containing x by [x]. That is for x ∈ X, [x] = {y ∈ X : yRx} is the equivalence class containing x. Now suppose x ∈ [x], which is true since xRx and also x ∈ [y], for some y ∈ X. What is [y]? Note that [y] = {z ∈ X : zRy}. If such a situation arises then we aim to prove that [x] = [y]. So let z ∈ [x]. This implies zRx. Also we have xRy. Now our relation R is a transitive relation. Hence zRy and xRy together implies that zRy this implies that z ∈ [y]. That is z ∈ [x] implies z ∈ [y] implies [x] ⊆ [y]. Exactly in the same way we can prove that [y] ⊆ [x]. Hence [x] = [y], whenever x ∈ [y].

For x, $y \in X$, either $[x]\cap[y]=\phi$ or [x] = [y]. For $x \in X$ there is exactly one equivalence class containing x. Such an equivalence class which we denoted by [x], is called a connected component (also known as a component or a maximal connected set containing x. Why should we call [x] as a maximal connected set containing x ? So we will have to prove that if E is a connected subset of X containing $x \in E$, then E ⊆ [x]. That is [x] is the largest connected set containing x. Have we proved that [x] is a connected subset of X?

Note that for each $y \in [x]$, there is a connected subset say E_y (E_y is just a notation) of X containing x, y. That is $\{E_y\}_{y\in[x]}$ is a collection of connected sets and $x \in E_y$ for all $y \in [x]$. This implies that $\bigcup_{y\in[x]} E_y$ is also connected set. It is simple exercise to see that $[x]=\bigcup_{y\in[x]} E_y$. Hence [$x$] is a connected set containing x. If E is a connected set containing x then for each $y \in E$, yRx. Hence $E \subseteq [x]$.

Also we know that if A is a connected subset of a topological space X then $\overline{A}$ is also a connected subset of X. Hence $\overline{[x]}$, the closure of the set [x], is also connected set containing x. But we have just proved that [x] is the maximal connected set containing x. Hence $\overline{[x]}\subseteq[x]$ implies $\overline{[x]}=[x]$ and [x] is a closed set. Therefore our maximal connected set containing x is a closed set.

Theorem 11. Intermediate value theorem. Let $(X, \mathcal{J})$ be a connected topological space and $f:(X, J) \to \mathbb{R}$ be a continuous function. If x, y are points of X and α is a real number such that α lies between f(x) and f(y). So let us say that f(x) < α < f(y). Then there exists z ∈ X such that f(z) = α.

Proof: Now f is a continuous function and X is a connected topological space implies *f(X)* is a connected subset of $\mathbb{R}$. But we know that every connected subset of $\mathbb{R}$ is an interval. Hence f(X) is an interval in $\mathbb{R}$. Now f(x), f(y) ∈ f(X) such that f(x) < f(y), f(X) is an interval implies [f(x), f(y)] ⊆ f(X). Suppose $\alpha \in \mathbb{R}$ such that f(x) < α < f(y).

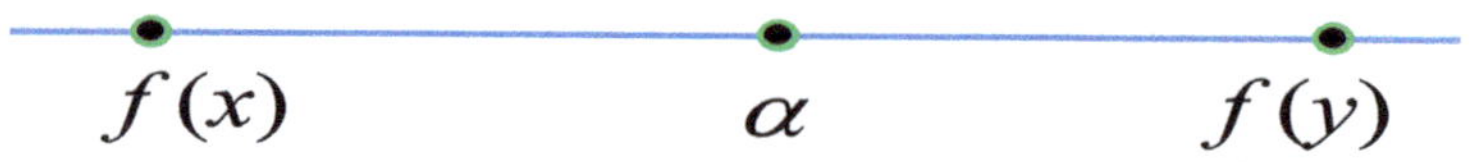

In particular α ∈ [f(x), f(y)] ⊆ f(X). That is α ∈ f(X). This implies that there exists z ∈ X such that f(z) = x.

Using the fact that $[a,b](a,b \in \mathbb{R},\ a<b)$ is a connected subspace of $\mathbb{R}$, we prove the following result:

Result: Let f: [*a*, *b*] → [*a*, *b*] be a continuous function, then there exists x_o in [*a*, *b*] such that $f(x_o) = x_o$.

Proof: (Proof by contradiction:)

Suppose $f(x) \neq x$ for each x ∈ [a, b]. Let A = {x ∈ [a, b] : f(x) < x }, B = {x ∈ [a, b] : f(x) > x}. Since *f* is a continuous function implies that both *A* and B are open subsets of [a, b]. Also $A \cap B = \phi$ and [a, b] = A ∪ B. Hence [a, b] is a connected topological space implies either $A = \phi$ or $B = \phi$.

Suppose $A = \phi$ then we get that B = [a, b]. That is [*a*, *b*] = {*x* ∈ [*a*, *b*] : f(x) > *x*}. In particular b ∈ [a, b] implies f(b) > b and this gives a contradiction to our assumption that f : [a, b] → [a, b]. (Note: If $B = \phi$ then A = [a, b] and a ∈ [a, b] implies f(a) < a again we will get a contradiction.) We get a contradiction if we assume that $f(x) \neq x$, for every x ∈ [*a*, *b*]. Hence there exists at least one x_o ∈ [*a*, *b*] such that $f(x_o) = x_o$.

Note. In the above result such an x_o ∈ [*a*, *b*] (satisfying $f(x_o) = x_o$) is called a fixed point of *f*.

Remark: In the proof of the above result we have used the fact that [a, b] is a connected topological subspace (in addition to the fact that f : [a, b] → [a, b] is a continuous map). What is to be noted here is we have not used the intermediate value theorem to prove the above result. Now let us use the intermediate value theorem to observe the following:

Define $g:[a,b] \to \mathbb{R}$ as g(x) = f(x) – x then g is a continuous map. Also g(a) = f(a) – a ≥ 0 and g(b) = f(b) – b ≤ 0. If f(a) = a or f(b) = b then we are through. If not g(b) = f(b)–b

$< 0 < f(a)-a = g(a)$. That is $g(b) < 0 < g(a)$. Hence by intermediate value theorem there exists $x_0 \in [a, b]$ such that $f(x_0) = x_0$.

Definition: A topological space $(X, \mathcal{J})$ is said to be totally disconnected if and only if the connected components of X are singletons. That is if A is a nonempty connected subset of X then A is a singleton set.

Exercises: Prove that $\mathbb{Q}$ and $\mathbb{R} \setminus \mathbb{Q} = \mathbb{Q}^c$ are totally disconnected topological spaces.

Now let us introduce a concept known as pathwise connected.

Definition: A topological space $(X, \mathcal{J})$ is said to be pathwise connected space if for x, $y \in X$ there exists a continuous function say $f : [0, 1] \to X$ such that $f(0) = x$, $f(1) = y$. That is, X is pathwise connected if and only if for $x, y \in X$ there exists a curve joining x and y.

Note. For $a, b \in \mathbb{R},\ a < b, [0,1]$ is homeomorphic to $[a, b]$. That is there exists a bijective continuous function $f : [0, 1] \to [a, b]$ such that $f^{-1} : [a, b] \to [0, 1]$ is also continuous. For let $f(t) = (1 - t)a + tb = a + (b - a)t$, $0 \le t \le 1$. Then f is bijective and f, f^{-1} are continuous functions.

Now let us prove the following:

Theorem 12. A topological space $(X, \mathcal{J})$ is pathwise connected if and only if for $x, y \in X$ and $a, b \in \mathbb{R},\ a < b$ there exists a continuous function $g : [a, b] \to X$ such that $g(a) = x$ and $g(b) = y$.

Proof: Given $x, y \in X$ and $a, b \in \mathbb{R},\ a < b$. Now X is a pathwise connected space implies there exists a continuous function say $f_1 : [0, 1] \to X$ such that $f_1(0) = x$ and $f_1(1) = y$. Now let $f_2(x) = \dfrac{x-a}{b-a},\ x \in [a,\ b]$. Now f_2 is a continuous function such that $f_2(a) = 0$ and $f_2(b) = 1$. Hence $g : f_1 \circ f_2 : [a, b] \to X$ is a continuous function such that $g(a) = f_1(f_2(a)) = f_1(0) = x$ and $g(b) = f_1(f_2(b)) = f_1(1) = y$.

Theorem 13. Every pathwise connected topological space X is a connected space.

Proof: Suppose X is disconnected. Then there exist nonempty closed subsets A, B of X such that (i) $X = A \cup B$ and (ii) $A \cap B = \phi$. Take $x \in A$, $y \in B$. Now X is a pathwise connected space implies there exists a continuous function say $f : [0, 1] \to X$ such that $f(0) = x$, $f(1) = y$. Hence $X = A \cup B$ implies $f^{-1}(X) = f^{-1}(A \cup B) = f^{-1}(A) \cup f^{-1}(B)$.

Also $x = f(0) \in A$ implies $0 \in f^{-1}(A)$, $y = f(1) \in B$ implies $1 \in f^{-1}(B)$ and $f^{-1}(A)$, $f^{-1}(B)$ are closed subsets of $[0, 1]$ ($f : [0, 1] \to X$ is continuous A, B are closed sets in X implies $f^{-1}(A)$, $f^{-1}(B)$ are closed sets in $[0, 1]$). Now $f^{-1}(A)$, $f^{-1}(B)$ are nonempty closed subsets of $[0,1]$ such that $[0, 1] = f^{-1}(X) = f^{-1}(A) \cup f^{-1}(B)$ and $f^{-1}(A) \cap f^{-1}(B) = f^{-1}(A \cap B) = \phi$ implies $[0, 1]$ is not connected and this gives a contradiction. Hence our initial assumption namely X is disconnected is not valid. Therefore X is a connected space.

It is easy to prove that the continuous image of a pathwise connected space is pathwise connected.

Result: If X is a nonempty convex subset of $\mathbb{R}^n (n \in \mathbb{N})$ then X is pathwise connected and hence X is a connected space.

Proof: For $x, y \in X$, define $f : [0, 1] \to X$ as $f(t) = (1 - t)x + ty$ for all $t \in [0, 1]$. Then for $t_n \in [0, 1]$, $t_n \to t \in [0, 1]$ implies $f(t_n) \to f(t)$. Hence f is a continuous function such that $f(0) = x$ and $f(1) = y$. Therefore X is pathwise connected.

The following result is known as pasting lemma is very useful in the study of connected spaces.

Lemma: (Pasting lemma) Let A, B be nonempty closed subsets of a topological space X. Suppose $f : A \to Y$, $g : B \to Y$ are continuous functions (where Y is a topological space) such that $f(x) = g(x)$, whenever $x \in A \cap B$. Then $h : A \cup B \to Y$ is defined as

$$h(x) = \begin{cases} f(x) & if\ x \in A \\ g(x) & if\ x \in B \end{cases}$$

is also a continuous function.

Proof: Left as an exercise to the reader. (Hint. For a nonempty closed subset C of Y, prove that $h^{-1}(C) = f^{-1}(C) \cup g^{-1}(C)$.)

Exercises: (i) In the above lemma assume that A, B are nonempty open sets and arrive at the conclusion that h defined as above is a continuous function.

(ii) Given that $(X, \mathcal{J})$ is a topological space. For $x, y \in X$ define xRy (read as x is related to y) if there is a path (or say curve) joining x and y. That is there exists a continuous function $f : [0, 1] \to X$ such that $f(0) = x$ and $f(1) = y$.

Hint. Now xRy implies there exists a continuous function $f : \left[0, \frac{1}{2}\right] \to X$ such that $f(0) = x$, $f\left(\frac{1}{2}\right) = y$. Also yRz implies there exists a continuous function $g : \left[\frac{1}{2}, 1\right] \to X$ such that $g\left[\frac{1}{2}\right] = y\ g(1) = z$. Now use pasting lemma to prove xRz. That is it is easy to prove that R is an equivalence relation on X.

Definition: A topological space $(X, \mathcal{J})$ is locally connected at a point x in X if and only if for each open set U containing x, there is a connected open set V such that $x \in V$ and $V \subseteq U$. If $(X, \mathcal{J})$ is locally connected at each x in X then we say that $(X, \mathcal{J})$ is a locally connected topological space.

Note that neither every connected topological space is a locally connected space nor every locally connected topological space is a connected space.

Example: Let $B=\left\{(x, y): 0<x\leq 1,\ y=sin\left(\frac{1}{x}\right)\right\}$ and $X=\overline{B}$. Define $f:(0,1]\rightarrow\mathbb{R}^2$ as $f(x)=\left(x, sin\left(\frac{1}{x}\right)\right)$ then f is continuous. Hence (0, 1] is a connected space implies the image f((0, 1]) = B is a connected subspace of $\mathbb{R}^2$. This implies that $\overline{B}=X$ is also a connected space and X = B ∪ (0 × [−1, 1]). This space X is called the topologist's sine curve and X is neither locally connected nor pathwise connected. Now $U=B\left(\left(0,\frac{1}{2}\right),\frac{1}{4}\right)\cap X$ is an open set containing $\left(0,\frac{1}{2}\right)$.

Now we leave it as an exercise to prove that there cannot exist any connected open set V satisfying $\left(0,\frac{1}{2}\right)\in V$ and V ⊆ U.

Example: X = (0, 1) ∪ (2, 3) is locally connected but it is not connected.

Theorem 14. Let Y be a connected open subset of $\mathbb{R}^n$ ($\mathbb{R}^n$ with Euclidean metric). Then Y is pathwise connected.

Proof: Fix $x\in Y$. Let us prove that for each y ∈ Y there exists a path joining x and y. So let A = {y ∈ Y : there is a path joining y and x}. As $x\in A, A\neq\phi$. We aim to prove that A = Y. Let y ∈ A. This implies y ∈ Y, and hence there exists r > 0 such that B(y, r) ⊆ Y. For z ∈ B(y, r). Define f : [0, 1] → Y as f(t) = (1 − t)z + ty. Then (1 − t)z + ty ∈ B(y, r), f is a continuous function satisfying the condition that f(0) = z and f(1) = y. So f is a path joining z and y. Also y ∈ A implies there exists a path joining y and x. That is there exists a continuous function say g : [0, 1] → Y such that g(0) = y, g(1) = x.

Now define h : [0, 1] → Y as h(t) = f(2t) if $0\leq t\leq\frac{1}{2}$ and h(t) = g(2t − 1) if $\frac{1}{2}\leq t\leq 1$. (Note: $0\leq t\leq\frac{1}{2}\Leftrightarrow 0\leq 2t\leq 1$ and $\frac{1}{2}\leq t\leq 1\Leftrightarrow 0\leq 2t-1\leq 1$.) Here t → f(2t) is a continuous function on $\left[0,\frac{1}{2}\right]$ and t → g(2t−1) is a continuous function on $\left[\frac{1}{2},1\right]$ and f(2t) = g(2t − 1) when $t=\frac{1}{2}$. Hence by pasting lemma h : [0, 1] → Y is a continuous function such that h(0) = z and h(1) = x. So we have a path h joining x and z and hence z ∈ A. This gives that B(y, r) ⊆ A and hence each point y of A is an interior point of A. This proves that A is an open set in the subspace Y. Now in the same way we can prove that Y \A is also an open set in Y. That is Y = A ∪ (Y\A), where A, Y\A are both open sets in Y. As $x\in A, A\neq\phi$. It is given that Y is connected and hence $Y\setminus A=\phi$. Therefore Y = A. Now A is pathwise connected and hence Y is pathwise connected. (For y, z ∈ A = Y, there is a path joining y and x also there is a path joining z and x. If there is path joining z and x then there is also a path joining x and z. Again there is a path joining y and x also a path joining x and z implies there is a path joining y and z.)

Example of a topological space which is connected but not pathwise connected.

Let $Y_1 = \{(0, y) \in \mathbb{R}^2 : -1 \le y \le 1\}$, $Y_2 = \left\{\left(x, sin\left(\frac{\pi}{x}\right)\right) \in \mathbb{R}^2 : 0 < x \le 1\right\}$ and $X = Y_1 \cup Y_2$.

Let $f : (0,1] \to Y_2$ defined as $f(x) = \left(x, \sin\left(\frac{\pi}{x}\right)\right)$ is a continuous function, $f(0,1] = Y_2$ (that is f is surjective) and (0,1] is a connected space implies the continuous image $f(0, 1] = Y_2$ is a connected space. Also $Y_2' = Y_1$ and hence $\overline{Y_2} = Y_1 \cup Y_2 = X$. Now Y_2 is a connected space implies $\overline{Y_2} = X$ is also a connected space. Now let us prove that X is not pathwise connected. Suppose there exists a continuous function say $g : [0, 1] \to X$ such that g(0) = (0, 1) (that is we want to prove that there is no path joining (1, 0) and a point of Y_2), Y_1 is a closed subset of $\mathbb{R}^2$ implies Y_1 is also a closed subset of X (X is a subspace of $\mathbb{R}^2$) and hence $g^{-1}(Y_1)$ is a closed subset of [0, 1]. Also $g(0) = (0, 1) \in Y_1$ implies $0 \in g^{-1}(Y_1)$. Now let us prove that $g^{-1}(Y_1)$ is also an open subset of [0,1]. So, let $t \in g^{-1}(Y_1)$ then $g(t) \in Y_1$.

Now $B\left(g(t), \frac{1}{2}\right) \cap X$ is an open set in X and g is a continuous function implies $g^{-1}\left(B\left(g(t), \frac{1}{2}\right) \cap X\right)$ is an open set in [0, 1]. Also $t \in g^{-1}\left(B\left(g(t), \frac{1}{2}\right) \cap X\right)$ implies there exists r > 0 such that $(t - r,\ t + r) \cap [0,\ 1] \subseteq g^{-1}\left(B\left(g(t), \frac{1}{2}\right) \cap X\right)$. Hence $g((t - r, t + r) \cap [0,1] \subseteq B\left(g(t), \frac{1}{2}\right) \cap X, B\left(g(t), \frac{1}{2}\right) \cap X$ consists of an interval on the y-axis, together with segments of the curve $y = sin\left(\frac{\pi}{x}\right)$, each of which is homomorphic to an interval. Further any two of these sets are separated from one another in $B\left(g(t), \frac{1}{2}\right) \cap Y_1$. Hence $B\left(g(t), \frac{1}{2}\right) \cap Y_1$ is a connected component of $B\left(g(t), \frac{1}{2}\right) \cap X$ containing g(t). Also $g(t - r, t + r) \subseteq B\left(g(t), \frac{1}{2}\right) \cap Y_1$. That is $(t - r, t + r) \subseteq g^{-1}\left(B\left(g(t), \frac{1}{2}\right) \cap Y_1\right) \subseteq g^{-1}(Y_1)$ This proves that $g^{-1}(Y_1)$ is also an open set. That is $g^{-1}(Y_1)$ is a nonempty set which is both open and closed in [0,1]. Hence [0,1] is a connected space implies $g^{-1}(Y_1) = [0, 1]$. That is $g([0, 1]) \subseteq g(g^{-1}(Y_1)) \subseteq Y_1$.

Therefore there cannot exist any continuous function g : [0, 1] → X such that g(0) = (0, 1). This proves that X is not pathwise connected.

Subspace Topology

In topology and related areas of mathematics, a subspace of a topological space *X* is a subset S of X which is equipped with a topology induced from that of X called the subspace topology (or the relative topology, or the induced topology, or the trace topology).

Definition

Given a topological space (X,τ) and a subset S of X, the subspace topology on S is defined by

$$\tau_S = \{S \cap U \mid U \in \tau\}$$

That is, a subset of S is open in the subspace topology if and only if it is the intersection of with an open set in (X,τ). If S is equipped with the subspace topology then it is a topological space in its own right, and is called a subspace of (X,τ). Subsets of topological spaces are usually assumed to be equipped with the subspace topology unless otherwise stated.

Alternatively we can define the subspace topology for a subset S of X as the coarsest topology for which the inclusion map

$$\iota : S \hookrightarrow X$$

is continuous.

More generally, suppose ι is an injection from a set S to a topological space X. Then the subspace topology on S is defined as the coarsest topology for which ι is continuous. The open sets in this topology are precisely the ones of the form $\iota^{-1}(U)$ for U open in X. S is then homeomorphic to its image in X (also with the subspace topology) and ι is called a topological embedding.

A subspace S is called an open subspace if the injection ι is an open map, i.e., if the forward image of an open set of S is open in X. Likewise it is called a closed subspace if the injection ι is a closed map.

Terminology

The distinction between a set and a topological space is often blurred notationally, for convenience, which can be a source of confusion when one first encounters these definitions. Thus, whenever S is a subset of X, and (X,τ) is a topological space, then the unadorned symbols "S" and "X" can often be used to refer both to S and X considered as two subsets of X, and also to (S,τ_S) and (X,τ) as the topological spaces, related as discussed above. So phrases such as " S an open subspace of X " are used to mean that (S,τ_S) is an open subspace of (X,τ), in the sense used below -- that is that: (i) $S\in\tau$; and (ii) S is considered to be endowed with the subspace topology.

Examples

In the following, $\mathbb{R}$ represents the real numbers with their usual topology.

- The subspace topology of the natural numbers, as a subspace of $\mathbb{R}$, is the discrete topology.
- The rational numbers $\mathbb{Q}$ considered as a subspace of $\mathbb{R}$ do not have the discrete topology (the point 0 for example is not an open set in $\mathbb{Q}$). If a and b are rational, then the intervals (a, b) and $[a, b]$ are respectively open and closed, but if a and b are irrational, then the set of all x with $a < x < b$ is both open and closed.
- The set [0,1] as a subspace of $\mathbb{R}$ is both open and closed, whereas as a subset of $\mathbb{R}$ it is only closed.
- As a subspace of $\mathbb{R}$, $[0, 1] \cup [2, 3]$ is composed of two disjoint *open* subsets (which happen also to be closed), and is therefore a disconnected space.
- Let $S = [0, 1)$ be a subspace of the real line $\mathbb{R}$. Then $[0, 1/2)$ is open in S but not in $\mathbb{R}$. Likewise $[1/2, 1)$ is closed in S but not in $\mathbb{R}$. S is both open and closed as a subset of itself but not as a subset of $\mathbb{R}$.

Properties

The subspace topology has the following characteristic property. Let Y be a subspace of X and let $i: Y \to X$ be the inclusion map. Then for any topological space Z a map $f: Z \to Y$ is continuous if and only if the composite map $i \circ f$ is continuous.

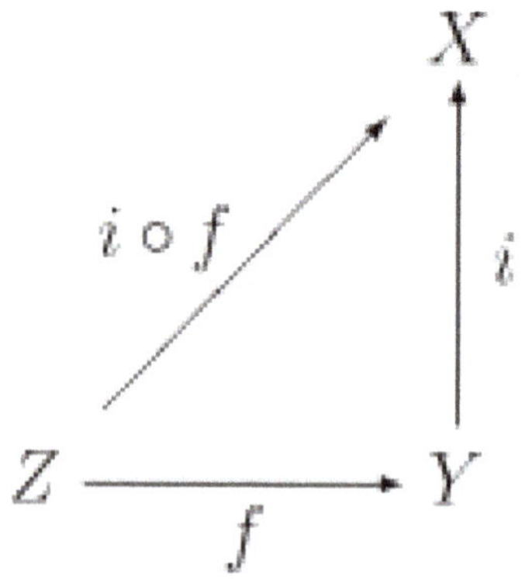

This property is characteristic in the sense that it can be used to define the subspace topology on Y.

We list some further properties of the subspace topology. In the following let S be a subspace of X.

- If $f: X \to Y$ is continuous the restriction to S is continuous.
- If $f: X \to Y$ is continuous then $f: X \to f(X)$ is continuous.
- The closed sets in S are precisely the intersections of S with closed sets in X.
- If A is a subspace of S then A is also a subspace of X with the same topology. In other words the subspace topology that A inherits from S is the same as the one it inherits from X.
- Suppose S is an open subspace of X (so $S \in \tau$). Then a subset of S is open in S if and only if it is open in X.
- Suppose S is a closed subspace of X (so $X \setminus S \in \tau$). Then a subset of S is closed in S if and only if it is closed in X.
- If B is a basis for X then $B_S = \{U \cap S : U \in B\}$ is a basis for S.
- The topology induced on a subset of a metric space by restricting the metric to this subset coincides with subspace topology for this subset.

Preservation of Topological Properties

If a topological space having some topological property implies its subspaces have that property, then we say the property is hereditary. If only closed subspaces must share the property we call it weakly hereditary.

- Every open and every closed subspace of a completely metrizable space is completely metrizable.
- Every open subspace of a Baire space is a Baire space.
- Every closed subspace of a compact space is compact.
- Being a Hausdorff space is hereditary.
- Being a normal space is weakly hereditary.
- Total boundedness is hereditary.
- Being totally disconnected is hereditary.
- First countability and second countability are hereditary.

Totally Disconnected Space

In topology and related branches of mathematics, a totally disconnected space is a topological space that is maximally disconnected, in the sense that it has no non-trivial connected subsets. In every topological space the empty set and the one-point sets are connected; in a totally disconnected space these are the *only* connected subsets.

An important example of a totally disconnected space is the Cantor set. Another example, playing a key role in algebraic number theory, is the field Q_p of *p*-adic numbers.

Definition

A topological space X is totally disconnected if the connected components in X are the one-point sets. Analogously, a topological space X is totally path-disconnected if all path-components in X are the one-point sets.

Examples

The following are examples of totally disconnected spaces:

- Discrete spaces
- The rational numbers
- The irrational numbers
- The p-adic numbers; more generally, all profinite groups are totally disconnected.
- The Cantor set and the Cantor space
- The Baire space
- The Sorgenfrey line
- Every Hausdorff space of small inductive dimension 0 is totally disconnected
- The Erdős space $\ell^2 \cap \mathbb{Q}^\omega$ is a totally disconnected Hausdorff space that does not have small inductive dimension 0.
- Extremally disconnected Hausdorff spaces
- Stone spaces
- The Knaster–Kuratowski fan provides an example of a connected space, such that the removal of a single point produces a totally disconnected space.

Properties

- Subspaces, products, and coproducts of totally disconnected spaces are totally disconnected.
- Totally disconnected spaces are T_1 spaces, since singletons are closed.

- Continuous images of totally disconnected spaces are not necessarily totally disconnected, in fact, every compact metric space is a continuous image of the Cantor set.
- A locally compact Hausdorff space has small inductive dimension 0 if and only if it is totally disconnected.
- Every totally disconnected compact metric space is homeomorphic to a subset of a countable product of discrete spaces.
- It is in general not true that every open set in a totally disconnected space is also closed.
- It is in general not true that the closure of every open set in a totally disconnected space is open, i.e. not every totally disconnected Hausdorff space is extremally disconnected.

Constructing a Disconnected Space

Let X be an arbitrary topological space. Let $x \sim y$ if and only if $y \in \text{conn}(x)$ (where $\text{conn}(x)$ denotes the largest connected subset containing x). This is obviously an equivalence relation whose equivalence classes are the connected components of X. Endow $X/\sim$ with the quotient topology, i.e. the finest topology making the map $m : x \mapsto \text{conn}(x)$ continuous. With a little bit of effort we can see that $X/\sim$ is totally disconnected. We also have the following universal property: if $f : X \to Y$ a continuous map to a totally disconnected space Y, then there exists a *unique* continuous map $\breve{f} : (X/\sim) \to Y$ with $f = \breve{f} \circ m$.

Union (Set Theory)

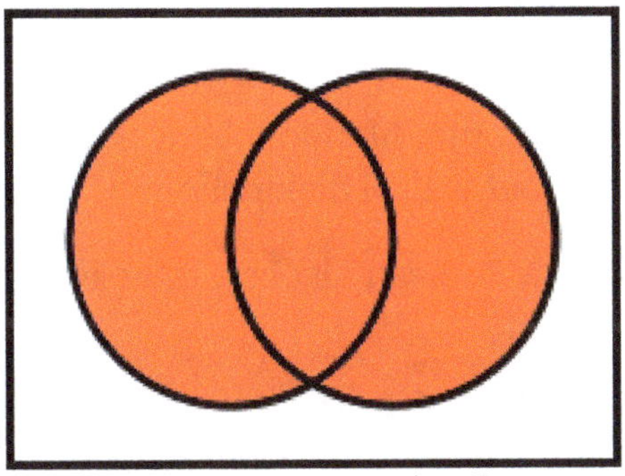

Union of two sets: $A \cup B$

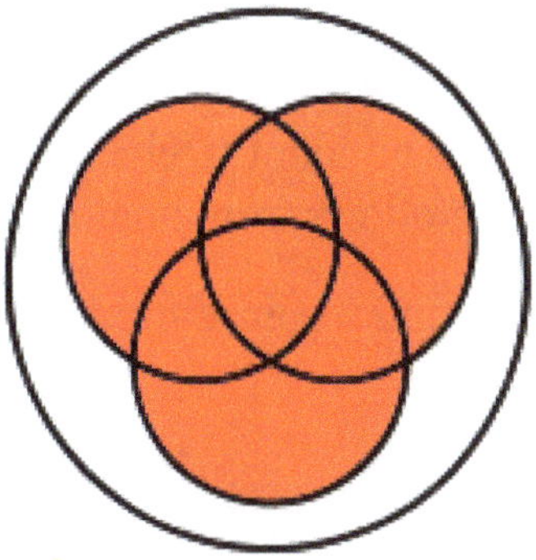

Union of three sets: $A \cup B \cup C$

In set theory, the union (denoted by ∪) of a collection of sets is the set of all elements in the collection. It is one of the fundamental operations through which sets can be combined and related to each other.

Union of Two Sets

The union of two sets A and B is the set of elements which are in A, in B, or in both A and B. In symbols,

$$A \cup B = \{x : x \in A \quad \text{or} \quad x \in B\}.$$

For example, if $A = \{1, 3, 5, 7\}$ and $B = \{1, 2, 4, 6\}$ then $A \cup B = \{1, 2, 3, 4, 5, 6, 7\}$. A more elaborate example (involving two infinite sets) is:

$A = \{x \text{ is an even integer larger than } 1\}$

$B = \{x \text{ is an odd integer larger than } 1\}$

$A \cup B = \{2, 3, 4, 5, 6, \ldots\}$

As another example, the number 9 is *not* contained in the union of the set of prime numbers {2, 3, 5, 7, 11, ...} and the set of even numbers {2, 4, 6, 8, 10, ...}, because 9 is neither prime nor even.

Sets cannot have duplicate elements, so the union of the sets {1, 2, 3} and {2, 3, 4} is {1, 2, 3, 4}. Multiple occurrences of identical elements have no effect on the cardinality of a set or its contents.

Algebraic Properties

Binary union is an associative operation; that is,

$$A \cup (B \cup C) = (A \cup B) \cup C.$$

The operations can be performed in any order, and the parentheses may be omitted without ambiguity (i.e., either of the above can be expressed equivalently as $A \cup B \cup C$). Similarly, union is commutative, so the sets can be written in any order.

The empty set is an identity element for the operation of union. That is, $A \cup \emptyset = A$, for any set A. This follows from analogous facts about logical disjunction.

Since sets with unions and intersections form a Boolean algebra, intersection distributes over union

$$A \cap (B \cup C) = (A \cap B) \cup (A \cap C)$$

and union distributes over intersection

$$A \cup (B \cap C) = (A \cup B) \cap (A \cup C).$$

Within a given universal set, union can be written in terms of the operations of intersection and complement as

$$A \cup B = \left(A^C \cap B^C\right)^C$$

where the superscript ^{C} denotes the complement with respect to the universal set.

Finite Unions

One can take the union of several sets simultaneously. For example, the union of three sets *A*, *B*, and *C* contains all elements of *A*, all elements of *B*, and all elements of *C*, and nothing else. Thus, *x* is an element of $A \cup B \cup C$ if and only if *x* is in at least one of *A*, *B*, and *C*.

In mathematics a finite union means any union carried out on a finite number of sets; it does not imply that the union set is a finite set.

Arbitrary Unions

The most general notion is the union of an arbitrary collection of sets, sometimes called an *infinitary union*. If M is a set whose elements are themselves sets, then *x* is an element of the union of M if and only if there is at least one element *A* of M such that *x* is an element of *A*. In symbols:

$$x \in \bigcup \mathrm{M} \Leftrightarrow \exists A \in \mathrm{M},\ x \in A.$$

This idea subsumes the preceding sections, in that (for example) $A \cup B \cup C$ is the union of the collection {*A*, *B*, *C*}. Also, if M is the empty collection, then the union of M is the empty set.

Notations

The notation for the general concept can vary considerably. For a finite union of sets $S_1, S_2, S_3, \ldots, S_n$ one often writes $S_1 \cup S_2 \cup S_3 \cup \ldots \cup S_n$ or $\bigcup_{i=1}^{n} S_i$. Various common notations for arbitrary unions include $\bigcup \mathrm{M}$, $\bigcup_{A \in \mathrm{M}} A$, and $\bigcup_{i \in I} A_i$, the last of which refers to the union of the collection $\{A_i : i \in I\}$ where *I* is an index set and A_i is a set for every $i \in I$. In the case that the index set *I* is the set of natural numbers, one uses a notation $\bigcup_{i=1}^{\infty} A_i$ analogous to that of the infinite series.

Whenever the symbol "∪" is placed before other symbols instead of between them, it is of a larger size.

Disjoint Sets

In mathematics, two sets are said to be disjoint if they have no element in common.

Equivalently, disjoint sets are sets whose intersection is the empty set. For example, {1, 2, 3} and {4, 5, 6} are disjoint sets, while {1, 2, 3} and {3, 4, 5} are not.

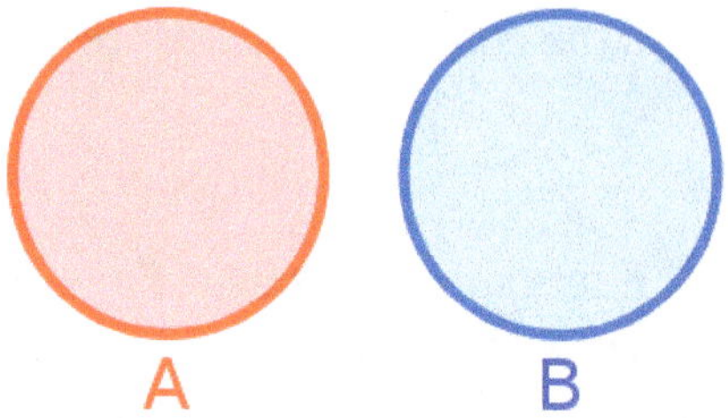

Two disjoint sets.

Generalizations

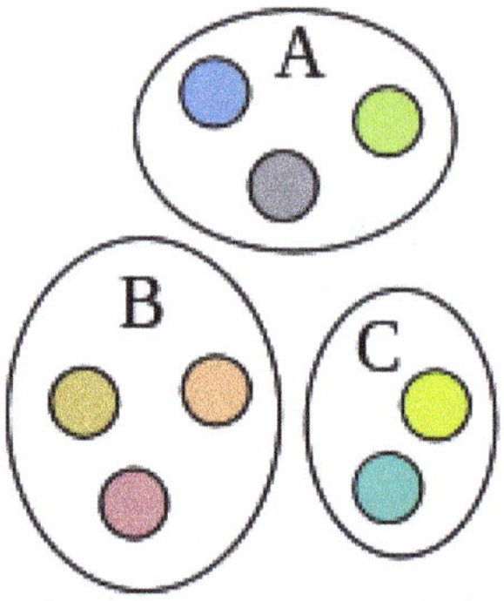

A pairwise disjoint family of sets

This definition of disjoint sets can be extended to any family of sets. A family of sets is pairwise disjoint or mutually disjoint if every two different sets in the family are disjoint. For example, the collection of sets { {1}, {2}, {3}, ... } is pairwise disjoint.

Two sets are said to be almost disjoint sets if their intersection is small in some sense. For instance, two infinite sets whose intersection is a finite set may be said to be almost disjoint.

In topology, there are various notions of separated sets with more strict conditions than disjointness. For instance, two sets may be considered to be separated when they have disjoint closures or disjoint neighborhoods. Similarly, in a metric space, positively separated sets are sets separated by a nonzero distance.

Examples

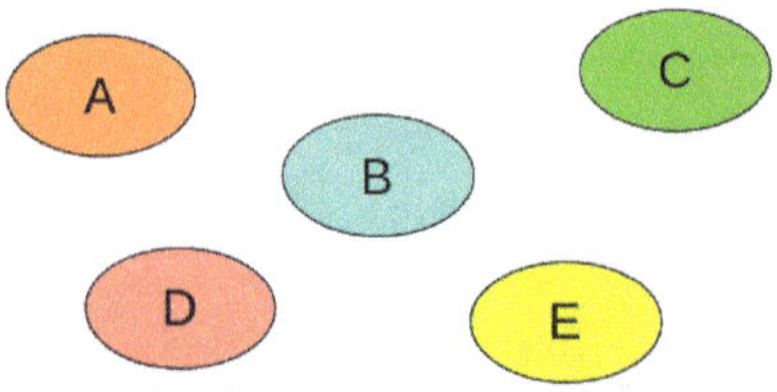

A pairwise disjoint family of sets

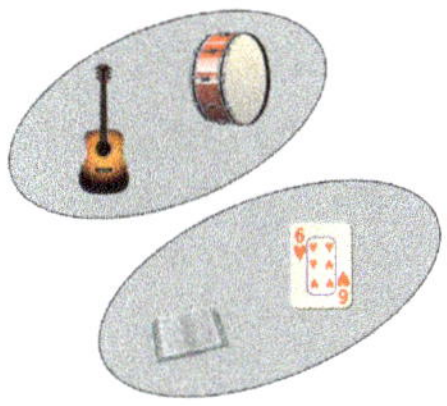

The set of the drum and the guitar is disjoint to the set of the card and the book

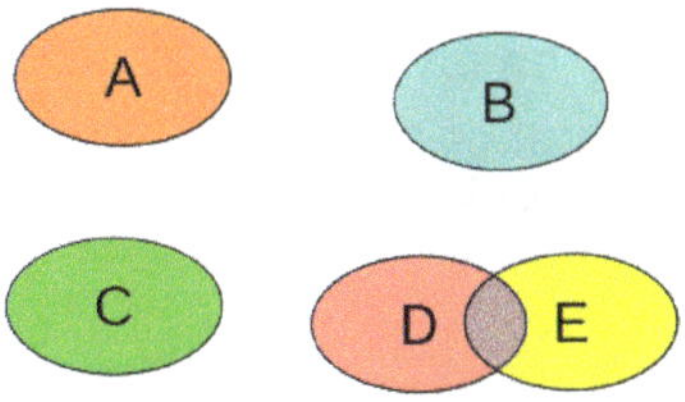

A non pairwise disjoint family of sets

Intersections

Disjointness of two sets, or of a family of sets, may be expressed in terms of their intersections.

Two sets A and B are disjoint if and only if their intersection $A \cap B$ is the empty set. It follows from this definition that every set is disjoint from the empty set, and that the empty set is the only set that is disjoint from itself.

A family F of sets is pairwise disjoint if, for every two sets in the family, their intersection is empty. If the family contains more than one set, this implies that the intersection of the whole family is also empty. However, a family of only one set is pairwise disjoint, regardless of whether that set is empty, and may have a non-empty intersection. Additionally, a family of sets may have an empty intersection without being pairwise disjoint. For instance, the three sets $\{ \{1, 2\}, \{2, 3\}, \{1, 3\} \}$ have an empty intersection but are not pairwise disjoint. In fact, there are no two disjoint sets in this collection. Also the empty family of sets is pairwise disjoint.

A Helly family is a system of sets within which the only subfamilies with empty intersections are the ones that are pairwise disjoint. For instance, the closed intervals of the real numbers form a Helly family: if a family of closed intervals has an empty intersection and is minimal (i.e. no subfamily of the family has an empty intersection), it must be pairwise disjoint.

Disjoint Unions and Partitions

A partition of a set X is any collection of mutually disjoint non-empty sets whose union is X. Every partition can equivalently be described by an equivalence relation, a binary relation that describes whether two elements belong to the same set in the partition. Disjoint-set data structures and partition refinement are two techniques in computer science for efficiently maintaining partitions of a set subject to, respectively, union operations that merge two sets or refinement operations that split one set into two.

A disjoint union may mean one of two things. Most simply, it may mean the union of sets that are disjoint. But if two or more sets are not already disjoint, their disjoint union may be formed by modifying the sets to make them disjoint before forming the union of the modified sets. For instance two sets may be made disjoint by replacing each element by an ordered pair of the element and a binary value indicating whether it belongs to the first or second set. For families of more than two sets, one may similarly replace each element by an ordered pair of the element and the index of the set that contains it.

Open Set

In topology, an open set is an abstract concept generalizing the idea of an open interval in the real line. The simplest example is in metric spaces, where open sets can be defined as those sets which contain an open ball around each of their points (or, equivalently, a set is open if it doesn't contain any of its boundary points); however, an open set, in general, can be very abstract: any collection of sets can be called open, as long as the union of an arbitrary number of open sets is open, the intersection of a finite number of open sets is open, and the space itself is open. These conditions are very loose, and they allow enormous flexibility in the choice of open sets. In the two extremes, every set can be open (called the discrete topology), or no set can be open but the space itself and the empty set (the indiscrete topology).

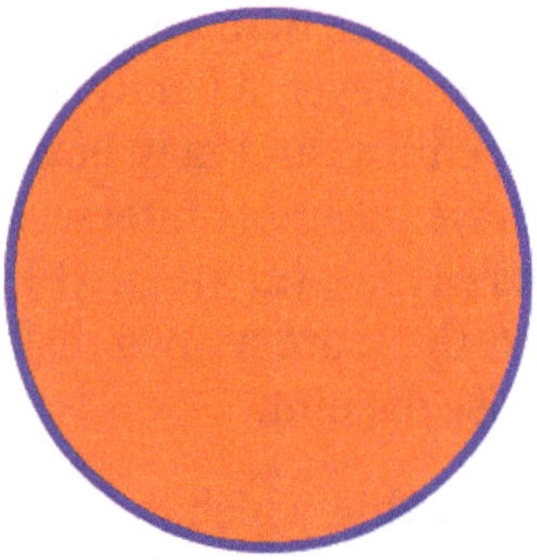

Example: The blue circle represents the set of points (x, y) satisfying $x^2 + y^2 = r^2$. The red disk represents the set of points (x, y) satisfying $x^2 + y^2 < r^2$. The red set is an open set, the blue set is its boundary set, and the union of the red and blue sets is a closed set.

In practice, however, open sets are usually chosen to be similar to the open intervals of the real line. The notion of an open set provides a fundamental way to speak of nearness of points in a topological space, without explicitly having a concept of distance defined. Once a choice of open sets is made, the properties of continuity, connectedness, and compactness, which use notions of nearness, can be defined using these open sets.

Each choice of open sets for a space is called a topology. Although open sets and the topologies that they comprise are of central importance in point-set topology, they are also used as an organizational tool in other important branches of mathematics. Examples of topologies include the Zariski topology in algebraic geometry that reflects the algebraic nature of varieties, and the topology on a differential manifold in differential topology where each point within the space is contained in an open set that is homeomorphic to an open ball in a finite-dimensional Euclidean space.

Motivation

Intuitively, an open set provides a method to distinguish two points. For example, if about one point in a topological space there exists an open set not containing another (distinct) point, the two points are referred to as topologically distinguishable. In this manner, one may speak of whether two subsets of a topological space are "near" without concretely defining a metric on the topological space. Therefore, topological spaces may be seen as a generalization of metric spaces.

In the set of all real numbers, one has the natural Euclidean metric; that is, a function which measures the distance between two real numbers: $d(x, y) = |x - y|$. Therefore, given a real number, one can speak of the set of all points close to that real number; that is, within ε of that real number (refer to this real number as x). In essence, points within ε of x approximate x to an accuracy of degree ε. Note that $\varepsilon > 0$ always but as ε becomes smaller and smaller, one obtains points that approximate x to a higher and higher degree of accuracy. For example, if $x = 0$ and $\varepsilon = 1$, the points within ε of x are precisely the points of the interval (-1, 1); that is, the set of all real numbers between -1 and 1. However, with $\varepsilon = 0.5$, the points within ε of x are precisely the points of (-0.5, 0.5). Clearly, these points approximate x to a greater degree of accuracy compared to when $\varepsilon = 1$.

The previous discussion shows, for the case $x = 0$, that one may approximate x to higher and higher degrees of accuracy by defining ε to be smaller and smaller. In particular, sets of the form $(-\varepsilon, \varepsilon)$ give us a lot of information about points close to $x = 0$. Thus, rather than speaking of a concrete Euclidean metric, one may use sets to describe points close to x. This innovative idea has far-reaching consequences; in particular, by defining different collections of sets containing 0 (distinct from the sets $(-\varepsilon, \varepsilon)$), one may find different results regarding the distance between 0 and other real numbers. For example, if we were to define R as the only such set for "measuring distance", all points are close to 0 since there is only one possible degree of accuracy one may achieve in approximating 0: being a member of R. Thus, we find that in some sense, every real number is distance 0 away from 0. It may help in this case to think of the measure as being a binary condition, all things in R are equally close to 0, while any item that is not in R is not close to 0.

In general, one refers to the family of sets containing 0, used to approximate 0, as a neighborhood basis; a member of this neighborhood basis is referred to as an open set. In fact, one may generalize these notions to an arbitrary set (X); rather than just the real numbers. In this case, given a point (x) of that set, one may define a collection of sets "around" (that is, containing) x, used to approximate x. Of course, this collection would have to satisfy certain properties (known as axioms) for otherwise we may not have a well-defined method to measure distance. For example, every point in X should approximate x to *some* degree of accuracy. Thus X should be in this family. Once we begin to define "smaller" sets containing x, we tend to approximate x to a greater degree of accuracy. Bearing this in mind, one may define the remaining axioms that the family of sets about x is required to satisfy.

Definitions

The concept of open sets can be formalized with various degrees of generality, for example:

Euclidean Space

A subset U of the Euclidean n-space R^n is called *open* if, given any point x in U, there exists a real number $\varepsilon > 0$ such that, given any point y in R^n whose Euclidean distance from x is smaller than ε, y also belongs to U. Equivalently, a subset U of R^n is open if every point in U has a neighborhood in R^n contained in U.

Metric Spaces

A subset U of a metric space (M, d) is called *open* if, given any point x in U, there exists a real number $\varepsilon > 0$ such that, given any point y in M with $d(x, y) < \varepsilon$, y also belongs to U. Equivalently, U is open if every point in U has a neighborhood contained in U.

This generalizes the Euclidean space example, since Euclidean space with the Euclidean distance is a metric space.

Topological Spaces

In general topological spaces, the open sets can be almost anything, with different choices giving different spaces.

Let X be a set and τ be a family of sets. We say that τ is a topology on X if:

- $X \in \tau, \varnothing \in \tau$ (X and $\varnothing$ are in τ)
- $\{O_i\}_{i\in I} \subseteq \tau \Rightarrow \cup_{i\in I} O_i \in \tau$ (any union of sets in τ is in τ)
- $\{O_i\}_{i=1}^{n} \subseteq \tau \Rightarrow \cap_{i=1}^{n} O_i \in \tau$ (any finite intersection of sets in τ is in τ)

We call the sets in τ the open sets.

Note that infinite intersections of open sets need not be open. For example, the intersection of all intervals of the form $(-1/n, 1/n)$, where n is a positive integer, is the set $\{0\}$ which is not open in the real line. Sets that can be constructed as the intersection of countably many open sets are denoted G_δ sets.

The topological definition of open sets generalizes the metric space definition: If one begins with a metric space and defines open sets as before, then the family of all open sets is a topology on the metric space. Every metric space is therefore, in a natural way, a topological space. There are, however, topological spaces that are not metric spaces.

Properties

The union of any number of open sets is open. The intersection of a finite number of open sets is open.

A complement of an open set (relative to the space that the topology is defined on) is called a closed set. A set may be both open and closed (a clopen set). The empty set and the full space are examples of sets that are both open and closed.

Uses

Open sets have a fundamental importance in topology. The concept is required to define and make sense of topological space and other topological structures that deal with the notions of closeness and convergence for spaces such as metric spaces and uniform spaces.

Every subset A of a topological space X contains a (possibly empty) open set; the largest such open set is called the interior of A. It can be constructed by taking the union of all the open sets contained in A.

Given topological spaces X and Y, a function f from X to Y is *continuous* if the preimage of every open set in Y is open in X. The function f is called *open* if the image of every open set in X is open in Y.

An open set on the real line has the characteristic property that it is a countable union of disjoint open intervals.

"Open" is Defined Relative to a Particular Topology

Whether a set is open depends on the topology under consideration. Having opted for greater brevity over greater clarity, we refer to a set X endowed with a topology T as "the topological space X" rather than "the topological space (X, T)", despite the fact that all the topological data is contained in T. If there are two topologies on the same set, a set U that is open in the first topology might fail to be open in the second topology. For example, if X is any topological space and Y is any subset of X, the set Y can be given its own topology (called the 'subspace topology') defined by "a set U is open in the subspace topology on Y if and only if U is the intersection of Y with an open set from the original topology on X." This potentially introduces new open sets: if V is open in the original topology on X, but $V \cap Y$ isn't, then $V \cap Y$ is open in the subspace topology on Y but not in the original topology on X.

As a concrete example of this, if U is defined as the set of rational numbers in the interval $(0, 1)$, then U is an open subset of the rational numbers, but not of the real numbers. This is because when the surrounding space is the rational numbers, for every point x in U, there exists a positive number a such that all *rational* points within distance a of x are also in U. On the other hand, when the surrounding space is the reals, then for every point x in U there is *no* positive a such that all *real* points within distance a of x are in U (since U contains no non-rational numbers).

Open and Closed are not Mutually Exclusive

A set might be open, closed, both, or neither.

For example, we'll use the real line with its usual topology (the Euclidean topology), which is defined as follows: every interval (a,b) of real numbers belongs to the topology, and every union of such intervals, e.g. $(a,b) \cup (c,d)$, belongs to the topology.

- In *any* topology, the entire set X is declared open by definition, as is the empty set. Moreover, the complement of the entire set X is the empty set; since X has an open complement, this means by definition that X is closed. Hence, in any topology, the entire space is simultaneously open and closed ("clopen").
- The interval $I = (0,1)$ is open because it belongs to the Euclidean topology. If I were to have an open complement, it would mean by definition that I were closed. But I does not have an open complement; its complement is $I^C = (-\infty, 0] \cup [1, \infty)$, which does *not* belong to the Euclidean topology since it is not a union of intervals of the form (a,b). Hence, I is an example of a set that is open but not closed.
- By a similar argument, the interval $J = [0,1]$ is closed but not open.
- Finally, since neither $K = [0,1)$ nor its complement $K^C = (-\infty, 0) \cup [1, \infty)$ belongs to the Euclidean topology (neither one can be written as a union of intervals of the form (a,b)), this means that K is neither open nor closed.

References

- George F. Simmons (1968). Introduction to Topology and Modern Analysis. McGraw Hill Book Company. p. 144. ISBN 0-89874-551-9
- Smith, Douglas; Eggen, Maurice; St. Andre, Richard (2010), A Transition to Advanced Mathematics, Cengage Learning, p. 95, ISBN 978-0-495-56202-3
- Paige, Robert; Tarjan, Robert E. (1987), "Three partition refinement algorithms", SIAM Journal on Computing, 16 (6): 973–989, MR 917035, doi:10.1137/0216062
- Lee, John M. (2010), Introduction to Topological Manifolds, Graduate Texts in Mathematics, 202 (2nd ed.), Springer, p. 64, ISBN 9781441979407
- Taylor, Joseph L. (2011). "Analytic functions". Complex Variables. The Sally Series. American Mathematical Society. p. 29. ISBN 9780821869017
- Juhász, István; Soukup, Lajos; Szentmiklóssy, Zoltán (2008). "Resolvability and monotone normality" (PDF). Israel Journal of Mathematics. The Hebrew University Magnes Press. 166 (1): 1–16. ISSN 0021-2172. doi:10.1007/s11856-008-1017-y. Retrieved 4 December 2012

4

Compact and Euclidean Space in Topology

Euclidean space includes the two-dimensional Euclidean plane and the three-dimensional Euclidean geometric plane and other spaces. Topological applications of Euclidean space is used to solve problems of Newtonian mechanics. The topics discussed in the chapter are of great importance to broaden the existing knowledge on topology.

Compact Space

In mathematics, and more specifically in general topology, compactness is a property that generalizes the notion of a subset of Euclidean space being closed (that is, containing all its limit points) and bounded (that is, having all its points lie within some fixed distance of each other). Examples include a closed interval, a rectangle, or a finite set of points. This notion is defined for more general topological spaces than Euclidean space in various ways.

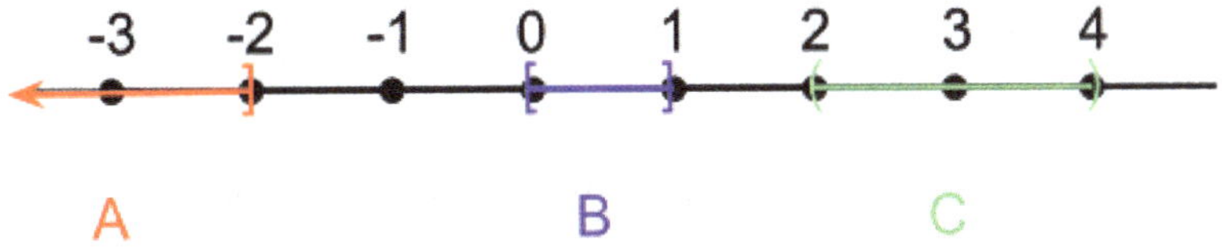

The interval A = (-∞, -2] is not compact because it is not bounded. The interval C = (2, 4) is not compact because it is not closed. The interval B = [0, 1] is compact because it is both closed and bounded.

One such generalization is that a space is *sequentially* compact if any infinite sequence of points sampled from the space must frequently (infinitely often) get arbitrarily close to some point of the space. An equivalent definition is that every sequence of points must have an infinite subsequence that converges to some point of the space. The Heine–Borel theorem states that a subset of Euclidean space is compact in this sequential sense if and only if it is closed and bounded. Thus, if one chooses an infinite number of points in the *closed* unit interval [0, 1] some of those points must get arbitrarily close to some real number in that space. For instance, some of the numbers 1/2, 4/5, 1/3, 5/6, 1/4, 6/7, ... accumulate to 0 (others accumulate to 1). The same set of points would not accumulate to any point of the *open* unit interval (0, 1); so the open unit interval is not compact. Euclidean space itself is not compact since it is not bounded. In particular, the sequence of points 0, 1, 2, 3, ... has no subsequence that converges to any given real number.

Apart from closed and bounded subsets of Euclidean space, typical examples of compact spaces include spaces consisting not of geometrical points but of functions. The term *compact* was introduced into mathematics by Maurice Fréchet in 1904 as a distillation of this concept. Compactness in this more general situation plays an extremely important role in mathematical analysis, because many classical and important theorems of 19th-century analysis, such as the extreme value theorem, are easily generalized to this situation. A typical application is furnished by the Arzelà–Ascoli theorem or the Peano existence theorem, in which one is able to conclude the existence of a function with some required properties as a limiting case of some more elementary construction.

Various equivalent notions of compactness, including sequential compactness and limit point compactness, can be developed in general metric spaces. In general topological spaces, however, different notions of compactness are not necessarily equivalent. The most useful notion, which is the standard definition of the unqualified term *compactness*, is phrased in terms of the existence of finite families of open sets that "cover" the space in the sense that each point of the space must lie in some set contained in the family. This more subtle notion, introduced by Pavel Alexandrov and Pavel Urysohn in 1929, exhibits compact spaces as generalizations of finite sets. In spaces that are compact in this sense, it is often possible to patch together information that holds locally—that is, in a neighborhood of each point—into corresponding statements that hold throughout the space, and many theorems are of this character.

The term compact set is sometimes a synonym for compact space, but usually refers to a compact subspace of a topological space.

Historical Development

In the 19th century, several disparate mathematical properties were understood that would later be seen as consequences of compactness. On the one hand, Bernard Bolzano (1817) had been aware that any bounded sequence of points (in the line or plane, for instance) has a subsequence that must eventually get arbitrarily close to some other point, called a limit point. Bolzano's proof relied on the method of bisection: the sequence was placed into an interval that was then divided into two equal parts, and a part containing infinitely many terms of the sequence was selected. The process could then be repeated by dividing the resulting smaller interval into smaller and smaller parts until it closes down on the desired limit point. The full significance of Bolzano's theorem, and its method of proof, would not emerge until almost 50 years later when it was rediscovered by Karl Weierstrass.

In the 1880s, it became clear that results similar to the Bolzano–Weierstrass theorem could be formulated for spaces of functions rather than just numbers or geometrical points. The idea of regarding functions as themselves points of a generalized space dates back to the investigations of Giulio Ascoli and Cesare Arzelà. The culmination of their investigations, the Arzelà–Ascoli theorem, was a generalization of the Bolzano–Weierstrass theorem to families of continuous functions, the precise conclusion of which was that it was possible to extract a uniformly convergent sequence of functions from a suitable family of functions. The uniform limit of this sequence then played precisely the same role as Bolzano's "limit point". Towards the beginning of the twentieth century, results similar to that of Arzelà and Ascoli began to accumulate in the area of integral equations, as investigated by David Hilbert and Erhard Schmidt. For a certain class of Green functions coming from solutions of integral equations, Schmidt had shown that a property analogous to the Arzelà–Ascoli theorem held in the sense of mean convergence—or convergence in what would later be dubbed a Hilbert space. This ultimately led to the notion of a compact operator as an offshoot of the general notion of a compact space. It was Maurice Fréchet who, in 1906, had distilled the essence of the Bolzano–Weierstrass property and coined the term *compactness* to refer to this general phenomenon (he used the term already in his 1904 paper which led to the famous 1906 thesis).

However, a different notion of compactness altogether had also slowly emerged at the end of the 19th century from the study of the continuum, which was seen as fundamental for the rigorous formulation of analysis. In 1870, Eduard Heine showed that a continuous function defined on a closed and bounded interval was in fact uniformly continuous. In the course of the proof, he made use of a lemma that from any countable cover of the interval by smaller open intervals, it was possible to select a finite number of these that also covered it. The significance of this lemma was recognized by Émile Borel (1895), and it was generalized to arbitrary collections of intervals by Pierre Cousin (1895) and Henri Lebesgue (1904). The Heine–Borel theorem, as the result is now known, is another special property possessed by closed and bounded sets of real numbers.

This property was significant because it allowed for the passage from local information about a set (such as the continuity of a function) to global information about the set (such as the uniform continuity of a function). This sentiment was expressed by Lebesgue (1904), who also exploited it in the development of the integral now bearing his name. Ultimately the Russian school of point-set topology, under the direction of Pavel Alexandrov and Pavel Urysohn, formulated Heine–Borel compactness in a way that could be applied to the modern notion of a topological space. Alexandrov & Urysohn (1929) showed that the earlier version of compactness due to Fréchet, now called (relative) sequential compactness, under appropriate conditions followed from the version of compactness that was formulated in terms of the existence of finite subcovers. It was this notion of compactness that became the dominant one, because it was not only a stronger property, but it could be formulated in a more general setting with a minimum of additional technical machinery, as it relied only on the structure of the open sets in a space.

Basic Examples

An example of a compact space is the (closed) unit interval [0,1] of real numbers. If one chooses an infinite number of distinct points in the unit interval, then there must be some accumulation point in that interval. For instance, the odd-numbered terms of the sequence 1, 1/2, 1/3, 3/4, 1/5, 5/6, 1/7, 7/8,... get arbitrarily close to 0, while the even-numbered ones get arbitrarily close to 1. The given example sequence shows the importance of including the boundary points of the interval, since the limit points must be in the space itself – an open (or half-open) interval of the real numbers is not compact. It is also crucial that the interval be bounded, since in the interval [0,∞) one could choose the sequence of points 0, 1, 2, 3,..., of which no sub-sequence ultimately gets arbitrarily close to any given real number.

In two dimensions, closed disks are compact since for any infinite number of points sampled from a disk, some subset of those points must get arbitrarily close either to a point within the disc, or to a point on the boundary. However, an open disk is not compact, because a sequence of points can tend to the boundary without getting arbitrarily close to any point in the interior. Likewise, spheres are compact, but a sphere missing a point is not since a sequence of points can tend to the missing point, thereby not getting arbitrarily close to any point *within* the space. Lines and planes are not compact, since one can take a set of equally-spaced points in any given direction without approaching any point.

Definitions

Various definitions of compactness may apply, depending on the level of generality. A subset of Euclidean space in particular is called compact if it is closed and bounded. This implies, by the Bolzano–Weierstrass theorem, that any infinite sequence from the set has a subsequence that converges to a point in the set. Various equivalent notions of compactness, such as sequential compactness and limit point compactness, can be developed in general metric spaces.

In general topological spaces, however, the different notions of compactness are not equivalent, and the most useful notion of compactness—originally called *bicompactness*—is defined using covers consisting of open sets. That this form of compactness holds for closed and bounded subsets of Euclidean space is known as the Heine–Borel theorem. Compactness, when defined in this manner, often allows one to take information that is known locally—in a neighbourhood of each point of the space—and to extend it to information that holds globally throughout the space. An example of this phenomenon is Dirichlet's theorem, to which it was originally applied by Heine, that a continuous function on a compact interval is uniformly continuous; here, continuity is a local property of the function, and uniform continuity the corresponding global property.

Open Cover Definition

Formally, a topological space X is called *compact* if each of its open covers has a finite subcover. Otherwise, it is called *non-compact*. Explicitly, this means that for every arbitrary collection

$$\{U_\alpha\}_{\alpha \in A}$$

of open subsets of X such that

$$X = \bigcup_{\alpha \in A} U_\alpha ,$$

there is a finite subset J of A such that

$$X = \bigcup_{i \in j} U_i .$$

Some branches of mathematics such as algebraic geometry, typically influenced by the French school of Bourbaki, use the term *quasi-compact* for the general notion, and reserve the term *compact* for topological spaces that are both Hausdorff and *quasi-compact*. A compact set is sometimes referred to as a *compactum*, plural *compacta*.

Compactness of Subspaces

A subset K of a topological space X is called compact if it is compact as a subspace. Explicitly, this means that for every arbitrary collection

$$\{U_\alpha\}_{\alpha \in A}$$

of open subsets of X such that

$$K \subseteq \bigcup_{\alpha \in A} U_\alpha ,$$

there is a finite subset J of A such that

$$K \subseteq \bigcup_{i \in J} U_i .$$

Compactness is a topological property: If $K \subset Z \subset Y$, with subspace Z equipped with the subspace topology, then K is compact in Z if and only if K is compact in Y.

Equivalent Definitions

Assuming the axiom of choice, the following are equivalent:

1. A topological space X is compact.
2. Every open cover of X has a finite subcover.
3. X has a sub-base such that every cover of the space by members of the sub-base has a finite subcover (Alexander's sub-base theorem)
4. Any collection of closed subsets of X with the finite intersection property has nonempty intersection.
5. Every net on X has a convergent subnet.
6. Every filter on X has a convergent refinement.
7. Every ultrafilter on X converges to at least one point.
8. Every infinite subset of X has a complete accumulation point.

Euclidean Space

For any subset A of Euclidean space R^n, A is compact if and only if it is closed and bounded; this is the Heine–Borel theorem.

As a Euclidean space is a metric space, the conditions in the next subsection also apply to all of its subsets. Of all of the equivalent conditions, it is in practice easiest to verify that a subset is closed and bounded, for example, for a closed interval or closed n-ball.

Metric Spaces

For any metric space (X,d), the following are equivalent:

1. (X,d) is compact.
2. (X,d) is complete and totally bounded (this is also equivalent to compactness for uniform spaces).
3. (X,d) is sequentially compact; that is, every sequence in X has a convergent subsequence whose limit is in X (this is also equivalent to compactness for first-countable uniform spaces).
4. (X,d) is limit point compact; that is, every infinite subset of X has at least one limit point in X.
5. (X,d) is an image of a continuous function from the Cantor set.

A compact metric space (X,d) also satisfies the following properties:

1. Lebesgue's number lemma: For every open cover of X, there exists a number $\delta > 0$ such that every subset of X of diameter $< \delta$ is contained in some member of the cover.
2. (X,d) is second-countable, separable and Lindelöf – these three conditions are equivalent for metric spaces. The converse is not true; e.g., a countable discrete space satisfies these three conditions, but is not compact.
3. X is closed and bounded (as a subset of any metric space whose restricted metric is d). The converse may fail for a non-Euclidean space; e.g. the real line equipped with the discrete metric is closed and bounded but not compact, as the collection of all singletons of the space is an open cover which admits no finite subcover. It is complete but not totally bounded.

Characterization by Continuous Functions

Let X be a topological space and C(X) the ring of real continuous functions on X. For each $p \in X$, the evaluation map

$$\mathrm{ev}_p : C(X) \to \mathrm{R}$$

given by $\mathrm{ev}_p(f)=f(p)$ is a ring homomorphism. The kernel of ev_p is a maximal ideal, since the residue field C(X)/ker ev_p is the field of real numbers, by the first isomorphism theorem. A topological space X is pseudocompact if and only if every maximal ideal in C(X) has residue field the real numbers. For completely regular spaces, this is equivalent to every maximal ideal being the kernel of an evaluation homomorphism. There are pseudocompact spaces that are not compact, though.

In general, for non-pseudocompact spaces there are always maximal ideals m in C(X) such that the residue field C(X)/m is a (non-archimedean) hyperreal field. The framework of non-standard analysis allows for the following alternative characterization of compactness: a topological space X is compact if and only if every point x of the natural extension $^*X$ is infinitely close to a point x_0 of X (more precisely, x is contained in the monad of x_0).

Hyperreal Definition

A space X is compact if its natural extension $^*X$ (for example, an ultrapower) has the property that every point of $^*X$ is infinitely close to a suitable point of $X \subset^* X$. For example, an open real interval X=(0,1) is not compact because its hyperreal extension *(0,1) contains infinitesimals, which are infinitely close to 0, which is not a point of X.

Properties of Compact Spaces

Functions and Compact Spaces

A continuous image of a compact space is compact. This implies the extreme value theorem: a continuous real-valued function on a nonempty compact space is bounded above and attains its supremum. (Slightly more generally, this is true for an upper semicontinuous function.) As a sort of converse to the above statements, the pre-image of a compact space under a proper map is compact.

Compact Spaces and Set Operations

A closed subset of a compact space is compact., and a finite union of compact sets is compact.

The product of any collection of compact spaces is compact. (This is Tychonoff's theorem, which is equivalent to the axiom of choice.)

Every topological space *X* is an open dense subspace of a compact space having at most one point more than *X*, by the Alexandroff one-point compactification. By the same construction, every locally compact Hausdorff space *X* is an open dense subspace of a compact Hausdorff space having at most one point more than *X*.

Ordered Compact Spaces

A nonempty compact subset of the real numbers has a greatest element and a least element.

Let *X* be a simply ordered set endowed with the order topology. Then *X* is compact if and only if *X* is a complete lattice (i.e. all subsets have suprema and infima).

Examples

- Any finite topological space, including the empty set, is compact. More generally, any space with a finite topology (only finitely many open sets) is compact; this includes in particular the trivial topology.
- Any space carrying the cofinite topology is compact.
- Any locally compact Hausdorff space can be turned into a compact space by adding a single point to it, by means of Alexandroff one-point compactification. The one-point compactification of R is homeomorphic to the circle S^1; the one-point compactification of R^2 is homeomorphic to the sphere S^2. Using the one-point compactification, one can also easily construct compact spaces which are not Hausdorff, by starting with a non-Hausdorff space.
- The right order topology or left order topology on any bounded totally ordered set is compact. In particular, Sierpiński space is compact.
- No discrete space with an infinite number of points is compact. The collection of all singletons of the space is an open cover which admits no finite subcover. Finite discrete spaces are compact.

- In R carrying the lower limit topology, no uncountable set is compact.
- In the cocountable topology on an uncountable set, no infinite set is compact. Like the previous example, the space as a whole is not locally compact but is still Lindelöf.
- The closed unit interval [0,1] is compact. This follows from the Heine–Borel theorem. The open interval (0,1) is not compact: the open cover

$$\left(\frac{1}{n},1-\frac{1}{n}\right)$$

for $n = 3,4,...$ does not have a finite subcover. Similarly, the set of *rational numbers* in the closed interval [0,1] is not compact: the sets of rational numbers in the intervals

$$\left[0,\frac{1}{\pi}-\frac{1}{n}\right] \text{and} \left[\frac{1}{\pi}+\frac{1}{n},1\right]$$

cover all the rationals in [0, 1] for $n = 4,5,...$ but this cover does not have a finite subcover. (Note that the sets are open in the subspace topology even though they are not open as subsets of R.)

- The set R of all real numbers is *not compact* as there is a cover of open intervals that does not have a finite subcover. For example, intervals $(n-1,n+1)$, where n takes all integer values in Z, cover R but there is no finite subcover.
- For every natural number n, the n-sphere is compact. Again from the Heine–Borel theorem, the closed unit ball of any finite-dimensional normed vector space is compact. This is not true for infinite dimensions; in fact, a normed vector space is finite-dimensional if and only if its closed unit ball is compact.
- On the other hand, the closed unit ball of the dual of a normed space is compact for the weak-* topology. (Alaoglu's theorem)
- The Cantor set is compact. In fact, every compact metric space is a continuous image of the Cantor set.
- Consider the set K of all functions f: R → [0,1] from the real number line to the closed unit interval, and define a topology on K so that a sequence $\{f_n\}$ in K converges towards $f \in K$ if and only if $\{f_n(x)\}$ converges towards $f(x)$ for all real numbers x. There is only one such topology; it is called the topology of pointwise convergence or the product topology. Then K is a compact topological space; this follows from the Tychonoff theorem.
- Consider the set K of all functions f: [0,1] → [0,1] satisfying the Lipschitz condition $|f(x) - f(y)| \leq |x - y|$ for all $x, y \in [0,1]$. Consider on K the metric induced by the uniform distance

$$d(f,g) = \sup_{x\in[0,1]} |f(x) - g(x)|$$

Then by Arzelà–Ascoli theorem the space K is compact.

- The spectrum of any bounded linear operator on a Banach space is a nonempty compact subset of the complex numbers C. Conversely, any compact subset of C arises in this manner, as the spectrum of some bounded linear operator. For instance, a diagonal operator on the Hilbert space ℓ^2 may have any compact nonempty subset of C as spectrum.

Algebraic Examples

- Compact groups such as an orthogonal group are compact, while groups such as a general linear group are not.
- Since the p-adic integers are homeomorphic to the Cantor set, they form a compact set.
- The spectrum of any commutative ring with the Zariski topology (that is, the set of all prime ideals) is compact, but never Hausdorff (except in trivial cases). In algebraic geometry, such topological spaces are examples of quasi-compact schemes, "quasi" referring to the non-Hausdorff nature of the topology.
- The spectrum of a Boolean algebra is compact, a fact which is part of the Stone representation theorem. Stone spaces, compact totally disconnected Hausdorff spaces, form the abstract framework in which these spectra are studied. Such spaces are also useful in the study of profinite groups.
- The structure space of a commutative unital Banach algebra is a compact Hausdorff space.
- The Hilbert cube is compact, again a consequence of Tychonoff's theorem.
- A profinite group (e.g., Galois group) is compact.

Compact Spaces and Related Results

Definition: A subset K of a topological space $(X, \mathcal{J})$ is said to be a compact set if $\mathcal{A}$ is a collection of open sets in X such that $K \subseteq \bigcup_{A\in\mathcal{A}} A$ then there exists $n \in \mathbb{N}$ and $A_1, A_2, A_3, \ldots, A_n \in \mathcal{A}$ such that $K \subseteq \bigcup_{i=1}^{n} A_i$. That is K is a compact subset of a topological space $(X, \mathcal{J})$ if and only if $\mathcal{A}$ is any open cover for K implies $\mathcal{A}$ has a finite subcollection say $\mathcal{A}_f$ that will also cover K.

Note. If $\mathcal{A}$ is a collection of open sets in $(X, \mathcal{J})$ and $K \subseteq X$ is such that $K \subseteq \bigcup_{A\in\mathcal{A}} A$, then we say that $\mathcal{A}$ is an open cover for K.

Example: Let X be nonempty set and $\mathcal{J}$ be a topology on X. Let K be a finite subset of X.

Case 1: $K = \phi$.

Then verify that K is a compact set (exercise).

Case 2: K is a nonempty finite set.

In this case, there exists $n \in \mathbb{N}$ and $x_1, x_2, \ldots, x_n \in X$ such that K = $\{x_1, x_2, \ldots, x_n\}$. Now suppose $\mathscr{A}$ is a collection of open sets in X and $\{x_1, x_2, \ldots, x_n\} = K \subseteq \bigcup_{A \in \mathscr{A}} A$. Then for each $i \in \{1, 2, \ldots, n\}$, $x_i \in A_i$ for some $A_i \in \mathscr{A}$. (Note that $i \neq j$ need not imply $A_i \neq A_j$.) Now $\mathscr{A}_f = \{A_1, A_2, \ldots, A_n\}$ is a finite subcollection of $\mathscr{A}$ such that $K = \{x_1, x_2, \ldots, x_n\} \subseteq A_1 \cup A_2 \cup \cdots \cup A_n = \bigcup_{A \in \mathscr{A}_f} A$. That is, we started with an open cover $\mathscr{A}$ for K and we could get a finite subcollection $\mathscr{A}_f$ of $\mathscr{A}$ that also covers K. Hence by the definition, K is a compact subset of $(X, \mathcal{J})$.

Example: Let X be a nonempty set and $\mathcal{J}_f = \{A \subseteq X : A^c = X \setminus A$ is a finite set or $A^c = X\}$. That is, $\mathcal{J}_f$ is the cofinite topology on X. We have proved in example that every finite subset of any topological space is compact.

So, let us assume that K is an infinite subset of X. Now consider a collection $\mathscr{A}$ of open sets such that $K \subseteq \bigcup_{A \in \mathscr{A}} A$. Since we have assumed K is an infinite set, $K \neq \phi$. Take an element say $x_0 \in K$. Now $x_0 \in K \subseteq \bigcup_{A \in \mathscr{A}} A$. This implies there exists $A_0 \in \mathscr{A}$ such that $x_0 \in A_0$. Now A_0 is a nonempty open set in the cofinite topological space $(X, \mathcal{J}_f)$ implies $X \setminus A_0 = A_0^c$ is a nonempty finite set (or $A_0^c = \phi \Rightarrow A_0 = X \Rightarrow \mathscr{A}_f = \{A_0\}$). Also $K \subseteq A_0 \Rightarrow \mathscr{A}_f = \{A_0\}$. Let $K \cap A_0^c = \{x_1, x_2, \ldots, x_n\}$. Since $K \subseteq \bigcup_{A \in \mathscr{A}} A$ each $x_i \in A_i$, $i = 1, 2, \ldots, n$. Now $K \subseteq X = A_0 \cup A_0^c \Rightarrow K \subseteq (A_0 \cup A_0^c) \cap K = (A_0 \cap K) \cup (A_0^c \cap K) \subseteq A_0 \cup A_1 \cup A_2 \cup \cdots \cup A_n$. That is, $\mathscr{A}_f = \{A_0, A_1, A_2, \ldots, A_n\}$ or $\mathscr{A}_f = \{A_0\}$ is a finite subcollection of $\mathscr{A}$ that also covers K. Hence K is a compact subset of $(X, \mathcal{J}_f)$. That is, in a cofinite topological space every subset is a compact set.

Remark: In a cofinite topological space $(X, \mathcal{J}_f)$, if A is a nonempty open set then A will almost cover any $K \subseteq X$. That is, maximum finitely many elements may not be in A and hence every subset K becomes a compact set.

Example: Let X be any set and $\mathcal{J}$ be a topology on X. Note that $\mathcal{J} \subseteq \mathcal{P}(X)$, the collection of all subsets of X. If $\mathcal{J}$ is a finite set then every subset K of X is compact in $(X, \mathcal{J})$.

Also note that in a cofinite topological space $(X, \mathcal{J}_f)$ every finite subset of X is closed and if $F \neq X$, F is not a finite set then F is not closed. Now let us prove that $\mathbb{R}$ with usual topology $\mathcal{J}_s$ is not a compact space. That is $(\mathbb{R}, \mathcal{J}_s)$ is not a compact space. Now we want to prove that the subset $\mathbb{R}$ of the topological space $(\mathbb{R}, \mathcal{J}_s)$ is not compact. Note that $\mathbb{R} \subseteq \bigcup_{n=1}^{\infty} (-n, n)$. That is, $\mathscr{A} = \{(-n, n) : n \in \mathbb{N}\}$ is an open cover for $\mathbb{R}$. Suppose this open cover has a finite subcollection say $\mathcal{A}_f = \{A_1, A_2, \ldots A_k\}$ such that $\mathbb{R} \subseteq \bigcup_{i=1}^{k} A_i$. If $A_i \in \mathcal{A}$ this implies there exists $n_i \in \mathbb{N}$ such that $A_i = (-n_i, n_i)$. So, $\mathbb{R} \subseteq (-n_1, n_1) \cup (-n_2, n_2) \cup \cdots \cup (-n_k, n_k)$. Let $n_o = \max\{n_1, n_2, \ldots, n_k\}$. Then $\mathbb{R} \subseteq (-n_0, n_0)$, a contradiction. Note that $n_0 + 1 \in \mathbb{R}$ but $n_0 + 1 \notin (-n_0, n_0)$. We could arrive at this contradiction by assuming that $\mathcal{A}$ has a finite subcollection that also covers $\mathbb{R}$. Hence such an assumption is wrong. That is, this particular collection $\mathcal{A} = \{(-n, n) : n \in \mathbb{N}\}$ is an open cover for $\mathbb{R}$. But this cannot have any finite sub-cover. Therefore $(\mathbb{R}, \mathcal{J}_s)$ is not a compact space. Note that $\mathbb{R}$ with cofinite topology $\mathcal{J}_f$ is a compact space. That is, $(\mathbb{R}, \mathcal{J}_f)$ is a compact topological space but $(\mathbb{R}, \mathcal{J}_s)$ is not a compact topological space.

Also note that if A is any unbounded subset of $\mathbb{R}$ then A is not a compact subset of the topological space $(\mathbb{R}, \mathcal{J}_s)$. Note that $A \subseteq \bigcup_{n=1}^{\infty} (-n, n)$ but A is not a bounded set (that is A is unbounded set) implies there cannot exist any $n_0 \in \mathbb{N}$ such that $A \subseteq (-n_o, n_o)$. (Recall: A is a bounded subset of $\mathbb{R}$ if and only if there exists $n_0 \in \mathbb{N}$ such that $|x| < n_o\ \forall x \in A$. That is A is bounded if and only if $A \subseteq (-n_o, n_o)$ for some $n_0 \in \mathbb{N}$.)

Now let us prove that a closed subset of a compact topological space is compact.

Theorem 1. If A is a closed subset of a compact topological space $(X, \mathcal{J})$ then A is a compact set in $(X, \mathcal{J})$.

Proof: Let $\mathcal{A}$ be a collection of open sets in X such that $A \subseteq \bigcup_{B \in \mathcal{A}} B$. Now $\mathcal{A}' = \mathcal{A} \cup \{A^c\}$ is a collection of open sets such that $X = A \cup A^c \subseteq A^c \cup \left(\bigcup_{B \in \mathcal{A}} B \right)$. That is, $\mathcal{A}'$ is an open cover for the compact space $(X, \mathcal{J})$ and hence there exists $n \in \mathbb{N}$ and $A_1, A_2, \ldots, A_n \in \mathcal{A}'$ such that $X \subseteq A_1 \cup A_2 \cup \cdots \cup A_n$. If one of $A_i = A^c$, then $\{A_1, A_2, \ldots, A_{i-1}, A_{i+1}, \ldots, A_n\}$ is a finite subcover of $\mathcal{A}$ that also covers A. If none of $A_i = A^c$ then $\{A_1, A_2, \ldots, A_n\} \subseteq \mathcal{A}$ such that $A \subseteq \bigcup_{i=1}^{n} A_i$. Hence in any case every open cover $\mathcal{A}$ of A has a finite subcollection that also covers A. This implies that A is a compact subset of $(X, \mathcal{J})$.

Recall that a topological space $(X, \mathcal{J})$ is said to be a Hausdorff topological space if $x, y \in X$, $x \neq y$ then there exist open sets U, V in X such that $x \in U$, $y \in V$ and $U \cap V = \phi$.

In a cofinite topological space we have proved that every subset is compact. In particular $(X, \mathcal{J}_f)$ is a compact topological space for any set X. But if $A \neq X$ is an infinite subset of X then A is a compact set but it is not a closed set. Now let us prove that such a thing cannot happen in a Hausdorff topological space.

Theorem 2. If K is a compact subset of a Hausdorff topological space then K is a closed set.

Proof: Let us prove that $K^c = X \backslash K$ is an open set. So take $x \in K^c$. Our aim is to prove that x is an interior point of K^c. That is, we will have to find an open set U_x in X such that $x \in U_x \subseteq K^c$. Now $x \neq y$ for each $y \in K$. Hence $(X, \mathcal{J})$ is a Hausdorff space implies there exist open sets U_y, V_y such that

$$x \in U_y,\ y \in V_y,\ U_y \cap V_y = \phi. \tag{1}$$

Now $\{V_y : y \in K\}$ is an open cover for the compact set K. Hence this implies there exist $y_1, y_2, \ldots, y_n \in K$ such that $K \subseteq \bigcup_{i=1}^{n} V_{y_i}$. Let $U_x = \bigcap_{i=1}^{n} U_{y_i}$ (refer Eq. (1)) then U_x is an open set containing x and $U_x \cap K \subseteq U_x \cap \left(\bigcup_{i=1}^{n} V_{y_i}\right) = \bigcup_{i=1}^{n}(U_x \cap V_{y_i}) \subseteq \bigcup_{i=1}^{n}(U_{y_i} \cap V_{y_i}) = \phi$. This implies $U_x \cap K = \phi \Rightarrow U_x \subseteq K^c$. Therefore, each $x \in K^c$ is an interior point of K^c. Hence K^c is an open set and therefore K is a closed set.

Note. Let $(X, \mathcal{J})$ be a topological space and $Y \subseteq X$. Then $\mathcal{J}_Y = \{A \cap Y : A \in \mathcal{J}\}$ is a topology on Y.

So, now it is easy to prove:

Theorem 3. A subset Y of a topological space $(X, \mathcal{J})$ is compact if and only if whenever $\mathcal{A}$ is a collection of open sets in $(Y, \mathcal{J}_Y)$ such that $Y = \bigcup_{A \in \mathcal{A}} A$ then there exists $n \in \mathbb{N}$ and $A_1, A_2, \ldots, A_n \in \mathcal{A}$ such that $Y = \bigcup_{i=1}^{n} A_i$.

Proof: Let us assume the given hypothesis. That is, assume that whenever $\mathcal{A}$ is a collection of open sets in the topological space $(Y, \mathcal{J}_Y)$ then this open cover $\mathcal{A}$ for Y has a finite subcover. Now we will have to prove that the given subset Y of the topological space $(X, \mathcal{J})$ is compact. So start with a collection say $\mathcal{B}$ of open sets in $(X, \mathcal{J})$ satisfying the condition that $Y \subseteq \bigcup_{B \in \mathcal{B}} B$ (recall the definition). Now $B \in \mathcal{B} \subseteq \mathcal{J} \Rightarrow B \cap Y \in \mathcal{J}_Y$. Hence $Y \subseteq \bigcup_{B \in \mathcal{B}} B$ implies $Y = \bigcup_{B \in \mathcal{B}} B \cap Y$. That is, $\mathcal{A} = \{B \cap Y : B \in \mathcal{B}\}$ is a collection of open sets in $(Y, \mathcal{J}_Y)$ which also covers Y. Hence by the given hypothesis there exists $n \in \mathbb{N}$ and $B_1, B_2, \ldots, B_n \in \mathcal{B}$ such that $Y = \bigcup_{i=1}^{n}(B_i \cap Y)$. This implies that $Y \subseteq \bigcup_{i=1}^{n} B_i$. Now we have proved: whenever $\mathcal{B}$ is a collection of open sets in $(X, \mathcal{J})$ which covers Y then there exists $n \in \mathbb{N}$ and $B_1, B_2, \ldots, B_n \in \mathcal{B}$ such that $Y \subseteq \bigcup_{i=1}^{n} B_i$. Therefore by our definition the given subset Y of $(X, \mathcal{J})$ is a compact set. The proof of Y is a compact subset of $(X, \mathcal{J})$ implies that the given hypothesis is satisfied follows in similar lines and hence the proof is left as an exercise.

From what we have proved, we observe that a subset Y of a topological space $(X, \mathcal{J})$ is compact if and only if, with respect to the induced topology $\mathcal{J}_Y$, the topological space $(Y, \mathcal{J}_Y)$ is compact.

Now let us prove that continuous image of a compact space is compact.

Theorem 4. Let $(X, \mathcal{J})$ be a compact topological space and $(Y, \mathcal{J}')$ be any other topological space. Let $f:(X, \mathcal{J}) \to (Y, \mathcal{J}')$ be a continuous function. Then the image *f(X)* is a compact subset of Y.

Proof: To prove the subset f(X) of $(Y, \mathcal{J}')$ is a compact set, we start with a collection say $\mathscr{A}$ of open sets in $(Y, \mathcal{J}')$ which satisfies

$$f(X) \subseteq \bigcup_{A \in \mathscr{A}} A. \qquad (2)$$

Now $A \in \mathscr{A}$ implies A is open in $(Y, \mathcal{J}')$. Hence $f:(X, \mathcal{J}) \to (Y, \mathcal{J}')$ is a continuous function implies that f^{-1} (A) is open in $(X, \mathcal{J})$. From Eq. (2) $X = f^{-1}(f(X)) \subseteq f^{-1}\left(\bigcup_{A \in \mathscr{A}} A\right) = \bigcup_{A \in \mathscr{A}} f^{-1}(A)$. This implies $\mathcal{A}' = \{f^{-1}(A) : A \in \mathscr{A}\}$ is an open cover for the compact topological space $(X, \mathcal{J})$. Hence there exists $n \in \mathbb{N}$ and $A_1, A_2, \ldots, A_n \in \mathscr{A}$ such that $X \subseteq \bigcup_{i=1}^{n} f^{-1}(A_i)$ (*here* $X = \bigcup_{i=1}^{n} f^{-1}(A_i)$). This implies that $f(X) = f\left(\bigcup_{i=1}^{n} f^{-1}(A_i)\right) = \bigcup_{i=1}^{n} f(f^{-1}(A_i)) \subseteq \bigcup_{i=1}^{n} A_i$. We have proved: any arbitrary open cover $\mathscr{A}$ of *f(X)* has a finite subcover. Hence by the definition, *f(X)* is a compact subset of $(Y, \mathcal{J}')$.

Using the above theorem and the result that every compact subset of a Hausdorff space is closed we prove:

Theorem 5. Let $(X, \mathcal{J})$ be a compact topological space and $(Y, \mathcal{J}')$ be a Hausdorff topological space. Let $f:(X, \mathcal{J}) \to (Y, \mathcal{J}')$ be a bijective continuous map. Then the inverse map $f^{-1}:(Y, \mathcal{J}') \to (X, \mathcal{J})$ is also a continuous map. That is f is a homeomorphism.

Proof: Take an open set A in $(X, \mathcal{J})$. Now *f* is a bijective map implies that

$$\left(f^{-1}\right)^{-1}\left(A^c\right) = f\left(A^c\right) \qquad (3)$$

Note that A^c is a closed subset of the compact space implies A^c is a compact set implies $f(A^c)$ is a compact subset of the Hausdorff space. This implies $f(A^c)$ is a closed set in Y. Hence from Eq. (3) $(f^{-1})^{-1}(A^c)$ is a closed set implies $f(A) = (f^{-1})^{-1}(A) = Y \setminus (f^{-1})^{-1}(A^c)$ is an open set.

We have proved: A is an open set in $(X, \mathcal{J})$ implies $(f^{-1})^{-1}(A)$ is an open set in Y. Hence $f^{-1}:(Y, \mathcal{J}') \to (X, \mathcal{J})$ is a continuous map.

Remark: To prove the above theorem it is also enough to prove that if B is a closed subset of X then f(B) is a closed subset of Y.

Definition: A collection $\mathcal{F}$ of subsets of a given set X is said to have finite intersection property (f.i.p) if for any $n \in \mathbb{N}$ and $F_1, F_2, \ldots, F_n \in \mathcal{F}$ then $\bigcap_{i=1}^{n} F_i \neq \phi$

Theorem 6. A topological space $(X, \mathcal{J})$ is a compact space if and only if whenever $\mathcal{F}$ is a collection of closed subsets of X which has f.i.p then $\bigcap_{F \in \mathcal{F}} F \neq \phi$.

Proof: Assume that $(X, \mathcal{J})$ is a compact topological space. Now start with a collection $\mathcal{F}$ of closed subsets of X which has the f.i.p. Our aim is to prove $\bigcap_{F \in \mathcal{F}} F \neq \phi$. To achieve this, let us use the method of proof by contradiction.

Suppose $\bigcap_{F \in \mathcal{F}} F \neq \phi$. Then by the DeMorgan's law $\left(\bigcap_{F \in \mathcal{F}} F \right)^c = \bigcup_{F \in \mathcal{F}} F^c = X$. This implies $\{F^c : F \in \mathcal{F}\}$ is an open cover for the compact space. Hence there exists $n \in \mathbb{N}$ and $F_1, F_2, \ldots, F_n \in \mathcal{F}$ such that $X = \bigcup_{i=1}^{n} F_i^c \Rightarrow X^c = \left(\bigcup_{i=1}^{n} F_i^c \right)^c = \bigcap_{i=1}^{n} F_i$. Therefore $\bigcap_{i=1}^{n} F_i = \phi$, a contradiction to the fact that $\mathcal{F}$ has the finite intersection property. We arrived at this contradiction by assuming that $\bigcap_{F \in \mathcal{F}} F = \phi$. Hence this is not a valid assumption. This implies $\bigcap_{F \in \mathcal{F}} F \neq \phi$. Let us leave the converse part as an exercise.

Now we prove that real valued continuous function on a compact topological space attains its maximum.

Theorem 7. Let $(X, \mathcal{J})$ be a compact topological space and $\mathcal{J}_s$ be the usual topology on $\mathbb{R}$. Let $f : (X, \mathcal{J}) \rightarrow (\mathbb{R}, \mathcal{J}_s)$ be a continuous function. Then there exists $x_0 \in X$ such that $f(x) \leq f(x_0)$ for all $x \in X$. That is f attains its maximum at x_0.

Proof: Let us use the method of proof by contradiction. Then for a given $a \in X$, f cannot attains its maximum at a. Hence there exists $a' \in X$ such that $f(a) < f(a')$. This means that $f(a) \in (-\infty, f(a'))$. (Fix any one $a' \in X$ satisfying $f(a) < f(a')$.) Hence $f(X) \subseteq \bigcup_{a \in X} (-\infty, f(a'))$.

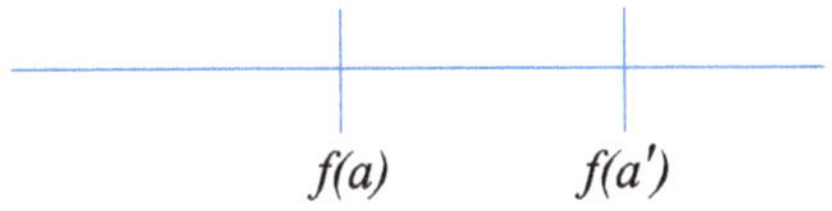

This implies that $\mathcal{A} = \{(-\infty, f(a')) : a \in X\}$ is an open cover for the compact subspace f(X) of $\mathbb{R}$ (continuous image of a compact space is compact). Hence there exist $a_1, a_2, \ldots, a_n \in X$ such that $f(X) \subseteq \bigcup_{i=1}^{n} (-\infty, f(a_i'))$. This implies $f(X) \subseteq (-\infty, f(a_0))$ for some $a_0 \in \{a_1', a_2', \ldots, a_n'\}$. Hence for this $a_0 \in X$ by our assumption there exists a $a' \in X$ such that $f(a_0) < f(a_0')$. But $f(X) \subseteq (-\infty, f(a_0))$ implies $f(a_0') < f(a_0)$. Hence we have got a contradiction. This means $f(a) < f(a')$ cannot be true for all $a \in X$ and hence there should exist at least one $x_0 \in X$ such that $f(x) \leq f(x_0)$ for all $x \in X$.

Remark: In a similar way we can prove that continuous image of a compact set attains its minimum at a point $y_o \in X$.

Theorem 8. Tychonoff. Let X and Y be compact topological spaces. Then the product space X × Y is compact.

Proof: For each $x_0 \in X$, $y \to (x_0, y)$ is a surjective continuous function and Y is a compact space implies $x_0 \times Y$ is a compact subset of X × Y. Let $\mathcal{A}$ be a collection of basic open sets such that $X \times Y = \bigcup_{U\times V \in \mathcal{A}} U \times V$.

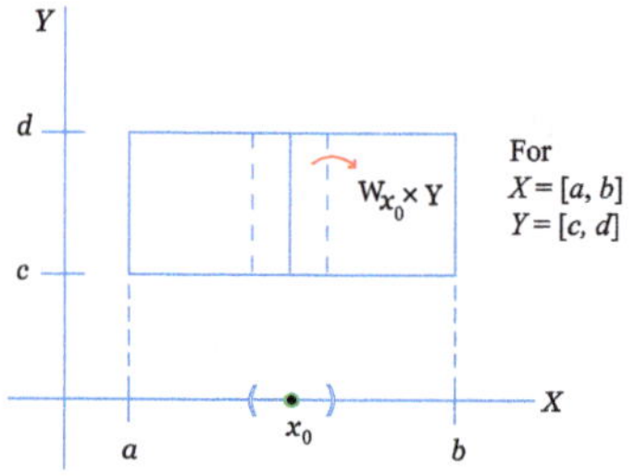

This implies that $x_0 \times Y \subseteq \bigcup_{U\times V \in \mathcal{A}} U \times V$ implies there exist $U_1\times V_1$, $U_2\times V_2$, $\cdots$, $U_n\times V_n$ such that

$$x_0 \times Y \subseteq (U_1 \times V_1) \cup (U_2 \times V_2) \cup \cdots \cup (U_n \times V_n). \qquad (4)$$

Also if for some i, $(U_i \times V_i) \cap (x_0 \times Y) = \phi$, then we do not require to include such an $U_i\times V_i$ in our finite subcover $\{U_i \times V_i\}_{i=1}^{n}$. So assume that each $(U_i \times V_i) \cap (x_0 \times Y) \neq \phi$. This in turn implies that $x_o \in U_i$, $\forall i = 1, 2, \ldots, n$ and hence $x_0 \in W_{x_0} = \bigcap_{i=1}^{n} U_i$. Now it is clear that $W_{x_0} \times Y \subseteq (U_1 \times V_1) \cup (U_2 \times V_2) \cup \cdots \cup (U_n \times V_n)$. Consider $(x, y) \in W_{x_0} \times Y$. Then $x \in U_i$ for all i and $y \in Y$. Hence from Eq. (4), $(x_o, y) \in U_j \times V_j$ for some j. This implies $(x, y) \in U_j \times V_j$ for the same j. That is for each $x_o \in X$, the tube $W_{x_0} \times Y$ is covered by finitely many members of $\mathcal{A}$.

Now let us prove that X × Y is covered by finitely many such tubes $W_x \times Y$. Now $\{W_x : x \in X\}$ is an open cover for X. Hence X is a compact space implies there exist $x_1, x_2, \ldots, x_k \in X$ such that $X = \bigcup_{i=1}^{k} W_{x_i}$. Now $(x, y) \in X \times Y \Rightarrow x \in W_{x_i}$ for some i, $1 \leq i \leq k$ and hence $(x, y) \in W_{x_i} \times Y$. This implies that $X \times Y \subseteq \bigcup_{i=1}^{k} W_{x_i} \times Y$ and hence X × Y is covered by finitely many members of $\mathcal{A}$. This proves that X × Y is a compact topological space.

Local Compactness

A Hausdorff topological space $(X, \mathcal{J})$ is said to be locally compact if and only if for each $x \in X$ and for each open set U containing x there exists an open set V containing x such that $\overline{V}$ is compact and $\overline{V} \subseteq U$. Now it is easy to prove that a Hausdorff topological space $(X, \mathcal{J})$ is locally compact if and only if for each $x \in X$ there exists an open set V such that $x \in V$ and $\overline{V}$ is a compact set in X.

Examples: (i) If X is a compact Hausdorff space then X is locally compact. (ii) $\mathbb{R}^n$ with Euclidean topology is locally compact but not compact. Here $n \in \mathbb{N}$ and J is the topology induced by the metric $d((x_1,x_2,\ldots,x_n),(y_1,y_2,\ldots,y_n)) = \left(\sum_{k=1}^{n}|x_k - y_k|^2\right)^{\frac{1}{2}}$.

Also it is easy to prove that if, for $1 \leq p \leq \infty$, $d_p((x_1,x_2,\ldots,x_n),(y_1,y_2,\ldots,y_n)) = \left(\sum_{k=1}^{n}|x_k - y_k|^p\right)^{\frac{1}{p}}$ and $d_\infty((x_1, x_2, \ldots, x_n),(y_1, y_2, \ldots, y_n)) = \max\{|x_k - y_k| : k = 1, 2, \ldots, n\}$ then d_p is a metric on $\mathbb{R}^n$. (Note. Proof of $d_p(x, y) \leq d_p(x, z) + d(z, y)$ for all $x, y, z \in \mathbb{R}^n$ is not that easy.)

For $x = (x_1, x_2, \ldots, x_n) \in \mathbb{R}^n$, $y = (y_1, y_2, \ldots, y_n) \in \mathbb{R}^n$, $\|x + y\| \leq \|x\| + \|y\|$ is known as Minkowski's inequality, $1 \leq p \leq \infty$.

If we use this inequality then $d_p(x, y) = \|x - y\|_p = \|(x - z) + (y - z)\|_p \leq \|x - z\|_p + \|z - y\|_p$ $= d_p(x,z) + d_p(z,y)$.

Also it is to be noted that the topology J_p on $\mathbb{R}^n$ induced by the metric d_p is same as $J_2 = J$.

Definition: A topological space (X, J) is said to be limit point compact if every infinite subset of X has a limit point.

Theorem 9. Every compact topological space (X, J) is limit point compact.

Proof: Let A be an infinite subset of X. Suppose $A' = \phi$. That is A does not have any limit point. Note that $\bar{A} = A \cup A' = A$ implies A is a closed set, then $x \in A$ implies $x \notin A'$ implies there exists an open set U_x such that $x \in U_x$, $U_x \cap A \setminus \{x\} = \phi$. (That is $U_x \cap A \cap \{x\}^c = \phi \Rightarrow U_x \cap A = \{x\}$. Now $\{U_x : x \in A\}$ is an open cover for the closed subset A of the given compact topological space. Hence there exists a natural number n and $x_1, x_2, \ldots, x_n \in A$ such that $A \subseteq U_{x_1} \cup \cdots \cup U_{x_n}$. This gives that $A = (U_{x_1} \cup \cdots \cup U_{x_n}) \cap A = (U_{x_1} \cap A) \cup (U_{x_2} \cap A) \cup \cdots \cup (U_{x_n} \cap A) = \{x_1, x_2, \ldots, x_n\}$. Hence we have arrived at a contradiction by assuming $A' = \phi$. Therefore $A' \neq \phi$. That is we have proved that every infinite subset A of the given compact topological space has at least one limit point. This means that (X, J) is a limit point compact topological space.

What about the converse of the above theorem? Is every limit point compact topological space compact? Limit point compact does not imply compact.

Example: Let $X = \{0,1\}$, $J = \{\phi, X\}$ and $Y = \mathbb{N} = \{1, 2, \ldots\}$, the set of all natural numbers and $J' = \mathcal{P}(\mathbb{N})$, that J' is discrete topology on $\mathbb{N}$. Let $X_0 = X \times Y$ be the product space. Here $\{X \times \{n\}\}$ is an open cover for $X \times Y$. But for any fixed $k \in \mathbb{N}$, $X \times Y = X \times \mathbb{N} \not\subseteq (X \times \{1\}) \cup \cdots \cup (X \times \{k\})$ (note: $(1, k+1) \notin \bigcup_{j=1}^{k} X \times \{j\}$). This gives that X × Y is not a compact topological space.

Now let A be a nonempty subset of X × Y . Then there exists $k \in \mathbb{N}$ such that (0, k) ∈ A or (1, k) ∈ A. Let us say (0, k) ∈ A. In this case we claim that $(1, k) \in A'$. Take a basic open set U containing (1, k) then U = X × {k}. Now $(0, k) \in U \cap A \setminus \{(1, k)\} \neq \phi$. Hence we have proved that (1, k) is a limit point of A. Note that if (1, k) ∈ A then we can prove that (0, k) is a limit point of A. So we have proved that every nonempty subset A of X × Y has a limit point. In particular every infinite subset of X × Y has a limit point. Therefore X × Y is a limit point compact.

One Point Compactification of a Topological Space

It is given that $(X, \mathcal{J})$ is a non compact Hausdorff topological space. Our aim is take an element say ∞ (just a notation) which is not in X. For each $x \in X$, we have open sets containing x. For ∞ ∈ $X^* = X \cup \{\infty\}$ we aim to define open sets satisfying: If U is an open set containing ∞ then each such open set is so large that the complement of U (with respect to X^*) is rather a small set. That is we want that $X^* \setminus U = C$, where C is a compact set in $(X, \mathcal{J})$ and since $\infty \in U$, $\infty \notin C$. So, if we start with a collection $\mathcal{A}$ of open sets in our new topological space (note: we have not yet defined such a topology on X^*) which covers X^*, then ∞ ∈ A_0 for some $A_0 \in \mathcal{A}$. Fix one such A_0. Now $X^* \setminus A_0$ = C a compact subset of $(X, \mathcal{J})$. So, if our proposed topology say $\mathcal{J}^*$ on X^* is such that $\mathcal{J}^*_X = \mathcal{J}$ then $\mathcal{A}$ is also an open cover for C and C is a compact subspace of $(X, \mathcal{J})$ implies C is also a compact subspace of $(X^*, \mathcal{J}^*)$. Hence there exists $n \in \mathbb{N}$ and A_1, $A_2, \ldots, A_n \in \mathcal{A}$ such that $C \subseteq A_1 \cup A_2 \cup \cdots \cup A_n$. Therefore $X^* = (X^* \setminus A) \cup A = C \cup A_o$ $\subseteq A_1 \cup \cdots \cup A_n \cup A_o$. Hence every open cover $\mathcal{A}$ of $(X^*, \mathcal{J}^*)$ has a finite subcover. So we see that if we could define such a topology $\mathcal{J}^*$ on X^* such that $\mathcal{J}^*_X = \mathcal{J}$ then $(X^*, \mathcal{J}^*)$ is a compact topological space.

We also want to retain the Hausdorff property. So keeping these points in mind we define $\mathcal{J}^*$ as follows:

- if a subset A of X^* is such that $\infty \notin A$ then $A \subseteq X$. In such a case $A \in \mathcal{J}^*$ if and only if $A \in \mathcal{J}$,
- if $\infty \in A$ then, $A \in \mathcal{J}^*$ if and only if $A = X^* \setminus C$, for some compact subset C of X.

Now it is easy to prove that $\mathcal{J}^*$ is a topology on X^*.

Theorem 10. Let $(X, \mathcal{J})$ be a locally compact Hausdorff space. Then there exists a topological space $(X^*, \mathcal{J}^*)$ satisfying the following conditions:

(i) $(X, \mathcal{J})$ is a subspace of $(X^*, \mathcal{J}^*)$,

(ii) $X^* \setminus X$ is a set containing exactly one element,

(iii) $(X^*, \mathcal{J}^*)$ is a compact Hausdorff space.

Proof: Keeping the above requirements in mind we have defined $\mathcal{J}^*$, we have taken care that $\mathcal{J} \subseteq \mathcal{J}^*$ and $\mathcal{J}^*_X = \mathcal{J}$. Also $X^* \setminus X = \{\infty\}$.

Now let us prove that $(X^*, \mathcal{J}^*)$ is a compact Hausdorff space. So start with a collection $\mathcal{A}$ of open sets in $(X^*, \mathcal{J}^*)$ (that is $\mathcal{A} \subseteq \mathcal{J}^*$) such that $X^* = \bigcup_{A \in \mathcal{A}} A$. Now $\infty \in X^*$ implies $\infty \in A_0$ for some $A_0 \in \mathcal{A}$. It is quite possible that ∞ belongs to more than one such $A \in \mathcal{A}$. From such A just fix one $A_0 \in \mathcal{A}$. By the definition of $\mathcal{J}^*$, $X^* \setminus A_0 = C$ is a compact subset of X. Note that $\mathcal{A}$ is a collection of open sets in X^* and C is a compact subspace of X, and hence of X^*. Now $C \subseteq \bigcup_{A \in \mathcal{A}} A$ implies there exists $n \in \mathbb{N}$ and $A_1, A_2, \ldots, A_n \in \mathcal{A}$ such that $C \subseteq A_1 \cup A_2 \cup \cdots \cup A_n$ implies $X^* = (X^* \setminus C) \cup C \subseteq A_0 \cup A_1 \cup \cdots \cup A_n$ ($X^* \setminus A_0 = C$ implies $A_0 = X^* \setminus C$) (it is possible that $A_0 = A_j$, for some $j \in \{1, 2, \ldots, n\}$. So we have proved that $X^* \subseteq A_0 \cup A_1 \cup \cdots \cup A_n$. This means that the started open cover $\mathcal{A}$ of X^* has finite subcover $\{A_0, A_1, A_2, \ldots, A_n\}$. Hence $(X^*, \mathcal{J}^*)$ is a compact space.

Now let us prove that $(X^*, \mathcal{J}^*)$ is a Hausdorff space. So start with $x, y \in X^*$ with $x \neq y$.

Case 1: $x, y \in X$ (means $x \neq \infty$, $y \neq \infty$).

Now $x, y \in X$, $(X, \mathcal{J})$ is a Hausdorff topological space implies there exist $U, V \in \mathcal{J}$ such that (i) $x \in U$, $y \in V$, (ii) $U \cap V = \phi$. But $\mathcal{J} \subseteq \mathcal{J}^*$. Hence we have $U, V \in \mathcal{J}^*$ satisfying (i) and (ii) and this is what we wanted to prove.

Case 2: $x \in X$, $y = \infty$.

Here we require the fact that $(X, \mathcal{J})$ is locally compact space. Now $x \in X$, $(X, \mathcal{J})$ is locally compact Hausdorff space implies there exists an open set U containing x such that $C = \bar{U}$ is a compact subset of $(X, \mathcal{J})$ (here $\bar{U}$ is the closure of U in X). Hence by the definition of $\mathcal{J}^*$, $V = X^* \setminus C$ is an open set containing ∞ and U is an open set containing x. Further $U \cap V = \phi$ and this is what we wanted to prove.

Remark: It is easy to prove that ∞ is a limit point of X. To prove this, start with an open set U containing ∞. Then we have to prove that $U \cap X \neq \phi$. If X is not a compact space, then $U \cap X \setminus \{\infty\} = U \cap X \neq \phi$. Hence if $(X, \mathcal{J})$ is not a compact Hausdorff space then ∞ is a limit point of X.

Examples: (i) Take $X = (a, b]$ and consider X as a subspace of $\mathbb{R}$ here $a, b \in \mathbb{R}$, $a < b$). Now X is considered as a subspace of $\mathbb{R}$, is a locally compact Hausdorff space of $\mathbb{R}$. Also X is not a compact space.

What is the one point compactification of X ? Here our $X = (a, b]$ and $a \notin X$. So take $\infty = a$. Note that while defining the one point compactification of X we just took an object or (say an element) which we denoted by ∞ and $\infty \notin X$. So what we need is

$\infty \notin X$. Now what are the open sets containing our ∞ = a in X^*. $U \subseteq X^*$ is an open set containing a if and only if $X^* \backslash U = C$ is a compact subset of X. Now it is easy to prove that $(X^*, \mathcal{J}^*)$ is homeomorphic to [a, b] as $f(x) = x$ for all $x \in X^*$. Now let us prove that f is a homeomorphism. Here it is enough to prove that f is a continuous map. So start with a nonempty open set U in [a, b] (here [a, b] is considered as a subspace of $\mathbb{R}$).

Case 1: $a \notin U$.

Then $U \subseteq (a, b]$. In this case by our definition U is open in X^*. That is $f^{-1}(U) = U$ is an open set in X^*.

Case 2: $a \in U$.

It is enough to consider a basic open set containing a. Hence $U = [a, a + \epsilon)$ for some $0 < \epsilon < b - a$. Is $f^{-1}(U) = f^{-1}([a, a + \epsilon))$ is an open set in $(X^*, \mathcal{J}^*)$. Note that U is an open set containing a if and only if $X^* \backslash U = [a, b] \backslash [a, a + \epsilon)$ is a compact subset of X(X = (a, b]). In our case $X^* \backslash U = [a + \epsilon, b]$ which is a compact subset of (a, b].

Hence from our definition of $\mathcal{J}^*$, $f^{-1}(U)$ is an open set in X^*. So from cases 1 and 2 we see that f is a continuous map. Now $f : (X^* \ \mathcal{J}^*) \to [a, b]$ such that

- f is bijective and continuous.
- $(X^*, \mathcal{J}^*)$ is a compact space.
- [a, b] is Hausdorff space implies f is a homeomorphism.

Hence we have proved that there is a homeomorphism between the one-point compactification of $(X^*, \mathcal{J}^*)$ and the compact Hausdorff space [a, b] and our X = (a, b] is such that X is proper subspace of [a, b] whose closure equals [a, b]. In such a case we say that [a, b] is a compactification of X.) So we define compactification of a topological space $(X, \mathcal{J})$ as follows:

Definition: A compact Hausdorff topological space $(Y, \mathcal{J}^*)$ is said to be a compactification of a topological space $(X, \mathcal{J})$ if and only if

(i) $(X, \mathcal{J})$ is a proper subspace of $(Y, \mathcal{J}^*)$,

(ii) $\overline{X} = Y$.

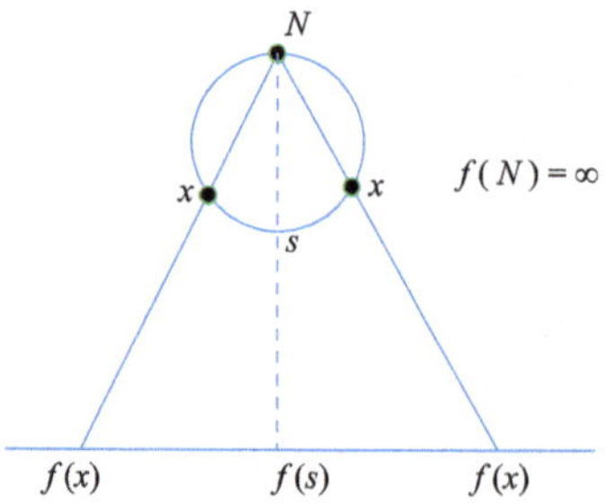

Note. If Y\X is a single point then we say that $(Y, \mathcal{J}^*)$ is the one point compactification of $(X, \mathcal{J})$. Now it is easy to prove the following statements:

(i) the one point compactification of $\mathbb{R}$ is homeomorphic to the unit circle $S^1 = \{(x_1, x_2) \in \mathbb{R}^2 : x_1^2 + x_2^2 = 1\}$,

(ii) the one point compactification of $\mathbb{R}^2$ is homeomorphic to the sphere $S^2 = \{(x_1, x_2, x_3) \in \mathbb{R}^2 : x_1^2 + x_2^2 + x_3^2 = 1\}$. So if we identify $\mathbb{R}^2$ with the complex plane (there is a homeomorphism between $\mathbb{R}^2$ and $\mathbb{C}$) then the one point compactification $\mathbb{C} \cup \{\infty\}$ of $\mathbb{C}$ is known as the Riemann sphere or the extended complex plane.

Tychonoff Theorem for Product Spaces

Now let us prove that if $(X_\alpha, \mathcal{J}_\alpha), \alpha \in J$ is an arbitrary collection of compact topological spaces then the product space $\prod_{\alpha \in J} X_\alpha$ is also a compact topological space. This theorem is due to Tychonoff and different proofs are available in the literature. To prove Tychonoff theorem we will use Zorn's lemma. Let us recall the following:

Definition: Let X be a nonempty set and $R \subseteq X \times X$, that is R is a relation on X. If $(x, y) \in R$ then we say that x is related to y and write $x \leq y$. The pair (X, R) is said to be a partially ordered set if and only if

(i) $x \leq x$ ($\leq$ is a reflexive),

(ii) for $x, y \in X, x \leq y$ and $y \leq x \Rightarrow x = y$. (That is $\leq$ is against symmetry in the sense that $x \leq y$ and $y \leq x$ can happen only when $x = y$.) In this case we say that $\leq$ is antisymmetry,

(iii) for $x, y, z \in X\, x \leq y$ and $y \leq z \Rightarrow x \leq z$. ($\leq$ is transitive.)

In this case we say that (X, $\leq$) is a partially ordered set (PO set).

Definition: Let (X, $\leq$) be a partially ordered set and A be a nonempty subset of X. Then an element $x \in X$ (note: x need not be in A) is called an upper bound of A if and only if $a \leq x$ for all a $\in$ A. An element $y \in Y$ is called a lower bound of A if and only if y $\leq$ a for all a $\in$ A. If there exists an $x_o \in X$ such that (i) x_o is an upper bound of A, (ii) $x \in X$ is an upper bound of A implies $x_o \leq x$ then such an upper bound x_o is called the least upper bound (lub) of A and we can easily show that l.u.b of A is unique, when it exists. An element $x_o \in X$ is called the greatest lower bound (glb) of A if it satisfies the following: (i) x_o is a lower bound of A, (ii) if $y_o \in X$ is a lower bound of A implies $y_o \leq x_o$.

Definition: An element $x_o \in X$ of a partially ordered set is called a maximal element of X if $x \in X$ is such that $x_o \leq x$ then $x = x_o$. An element $y_0 \in X$ is called a minimal element of X if $y \in X$ is such that $y \leq y_0$ then $y = y_0$.

Example: Let X = {1, 2, 3, 4, 5}, R = {(1, 2),(3, 4),(n, n) : n ∈ {1, 2, 3, 4, 5}}. If $(x, y) \in R$ then we say that $x \le y$. Here 2, 4, 5 ∈ X and they are maximal elements of X. Note that $(2,3) \notin R$ and hence 2 is not related to 3. That is 2 ≤ 3 is not true. Similarly 2 is not related to 4 and 2 is not related to 5. So 2 is not smaller than other elements of X and hence 2 is a maximal element of X. Since 3 ≤ 4 and $3 \neq 4$, 3 is not maximal element of X. If $y_0 \in X$ is such that y_0 is not larger than any other element of X then we say that y_0 is a minimal element of X. That is if there exists y ∈ X such that $y \le y_0$ then $y = y_0$.

A nonempty subset A of X is said to be a chain (also known as totally ordered set) if for $x, y \in A$, $x \le y$ or $y \le x$. That is any pair of elements x, y in A are comparable.

Now we are in a position to state Zorn's lemma:

Lemma: Zorn's Lemma. Let (X, ≤) be a partially ordered set. Further suppose every chain C ⊆ X has an upper bound in X. Then X will have at least one maximal element.

We observe the following: A topological space $(X, \mathcal{J})$ is compact if and only if whenever $\mathcal{A}$ is a collection of subsets of X which has finite intersection property (f.i.p) then $\bigcap_{A \in \mathcal{A}} \bar{A} \neq \phi$.

Theorem 11. Tychonoff theorem. Let $(X_\alpha, \mathcal{J}_\alpha), \alpha \in J$ be a collection of compact topological spaces. Then the product space $\left(\prod_{\alpha \in J} X_\alpha, \mathcal{J}\right)$ is also a compact space.

Proof: Start with a collection $\mathcal{A}$ of subsets of $X = \prod_{\alpha \in J} X_\alpha$ which has f.i.p. Then we aim to prove that $\bigcap_{A \in \mathcal{A}} \bar{A} \neq \phi$.

Step 1:

Let $\mathcal{F} = \{ \mathcal{D} : \mathcal{D}$ is a collection of subsets of X containing $\mathcal{A}$ and $\mathcal{D}$ has f.i.p $\}$.

For $\mathcal{D}_1, \mathcal{D}_2 \in \mathcal{F}$, define $\mathcal{D}_1 \le \mathcal{D}_2$ if $\mathcal{D}_1 \subseteq \mathcal{D}_2$. Then $(\mathcal{F}, \le)$ is a partially ordered set. Now let $\mathcal{C}$ be a chain in $\mathcal{F}$ and $\mathcal{A}_0 = \bigcup_{\mathcal{D} \in \mathcal{C}} \mathcal{D}$ (here $\mathcal{C} \subseteq \mathcal{F}$ and $\mathcal{D} \in \mathcal{F}$). It is easy to prove that $\mathcal{A}_0$ is an upper bound for $\mathcal{C}$. For this, we will have to prove that $\mathcal{A}_0 \in \mathcal{F}$ and $\mathcal{D} \le \mathcal{A}_0$ for all $\mathcal{D} \in \mathcal{C}$. First let us prove that $\mathcal{A}_0$ has f.i.p. Let $A_j \in \mathcal{A}_0$ for j = 1, 2, . . . , n. Then $A_j \in \mathcal{D}_j$, for some $\mathcal{D}_1, \mathcal{D}_2, \ldots, \mathcal{D}_n \in \mathcal{C}$. As $\mathcal{C}$ is a chain for j ∈ {1, 2, . . . , n} either $\mathcal{D}_i \subseteq \mathcal{D}_j$ or $\mathcal{D}_j \subseteq \mathcal{D}_i$. Hence there exists k, 1 ≤ k ≤ n such that $\mathcal{D}_j \subseteq \mathcal{D}_k$ for all j ∈ {1, 2, . . . , n}. Then $A_j \in \mathcal{D}_k$ for all j and $\mathcal{D}_k$ has f.i.p implies $\bigcap_{j=1}^{n} A_j \neq \phi$. Also $\mathcal{A} \subseteq \mathcal{A}_0$. Hence $\mathcal{A}_0 \subseteq \mathcal{F}$. By the definition of $\mathcal{A}_0$, $\mathcal{D} \subseteq \mathcal{A}_0$ for all $\mathcal{D} \subseteq \mathcal{C}$. This proves that $\mathcal{A}_0 \in \mathcal{F}$ is an upper bound for $\mathcal{C}$.

Now we have proved that every chain $\mathcal{C}$ in $\mathcal{F}$ has an upper bound in $\mathcal{F}$. Therefore by Zorn's lemma $\mathcal{F}$ will have a maximal element say $\mathcal{B} \in \mathcal{F}$. This $\mathcal{B} \in \mathcal{F}$ is such that

(i) $\mathcal{A} \subseteq \mathcal{B}, \mathcal{B}$ has f.i.p, (ii) whenever $\mathcal{A}'$ is a collection of subsets of X such that $\mathcal{A} \subseteq \mathcal{A}', \mathcal{A}'$ has f.i.p then $\mathcal{A}' \subseteq \mathcal{B}$.

Step 2: Now let us prove that $\mathcal{B}$ has the following properties:

(i) For $n \in \mathbb{N}$, $A_1, A_2, \ldots, A_n \in \mathcal{B}$ implies $A_1 \cap A_2 \cap \cdots \cap A_n \in \mathcal{B}$.

(ii) If A is subset of X such that $A \cap B \neq \phi$, for all $B \in \mathcal{B}$ then $A \in \mathcal{B}$.

To prove (i), let $A_0 = A_1 \cap A_2 \cap \cdots \cap A_n$ and $\mathcal{B}_0 = \mathcal{B} \cup \{A_0\}$. Then $\mathcal{B}_0 \in \mathcal{F}$ and $B \subseteq \mathcal{B}_0$. Since B is maximal, $B = \mathcal{B}_0$. This proves that $A_0 \in \mathcal{B}$.

To prove (ii), take $\mathcal{B}_0 = \mathcal{B} \cup \{A\}$. Then $\mathcal{B}_0 \in F$ and hence by step 1, $A \in \mathcal{B}$.

Step 3: Let us prove that $\bigcap_{A=\mathcal{B}} \overline{A} \neq \phi$.

For each $\alpha \in J$, $\{P_\alpha(A) : A \in \mathcal{B}\}$ is a collection of subsets of (X_α, J_α). If $A_1, A_2, \ldots, A_n \in \mathcal{B}$, then $\mathcal{B}$ has f.i.p and $\bigcap_{j=1}^{n} A_j \neq \phi$. Let $\bigcap_{i=1}^{n} A_j$. Now $P_\alpha(x) \in P_\alpha(A_j)$ for all $j = 1, 2, \ldots, n$. Hence $\{P_\alpha(A) : A \in \mathcal{B}\}$ is a collection of subsets of the compact topological space (X_α, J_α). Further this collection has f.i.p. This gives that $\bigcap_{A=\mathcal{B}} \overline{P_\alpha(A)} \neq \phi$. Let $x_\alpha \in \bigcap_{A=\mathcal{B}} \overline{P_\alpha(A)}$ and $x = (x_\alpha)_{\alpha \in J}$. (That is, we define $f : J \to \bigcup_{\alpha \in J} X_\alpha$ as $f(\alpha) = x_\alpha \in X_\alpha$ and we identify f with x.) Now we aim to prove that $x \in \overline{A}$, for each $A \in \mathcal{B}$. So fix $A \in \mathcal{B}$ and let $P_\beta^{-1}(V_\beta)$ be a subbasic open set containing x. Now $x = (x_\alpha) \in P_\beta^{-1}(V_\beta)$ implies $x_\beta \in V_\beta$. We have $x_\beta \in \overline{P_\beta(A)}$ and hence V_β is an open set in (X_α, J_α) containing x_β implies $V_\beta \cap P_\beta(A) \neq \phi$ implies there exists $y \in A$ such that $P_\beta(y) \in V_\beta$. This gives that $y \in P_\beta^{-1}(V_\beta) \cap A$. Hence $P_\beta\ (V_\beta) \cap A \neq$ for all $A \in \mathcal{B}$ implies $P_\beta^{-1}(V_\beta) \in \mathcal{B}$. Again if B is a basic open set containing x in the product space (X, J) then $B = P_{\beta_1}^{-1}(V_{\beta_1}) \cap P_{\beta_2}^{-1}(V_{\beta_2}) \cap \cdots \cap P_{\beta_n}^{-1}(V_{\beta_n})$ for some $V_{\beta_i} \in J_{\beta_i}, i = 1, 2, 3, \ldots, n$. We have proved that each $P_{\beta_i}^{-1}(V_{\beta_i}) \in \mathcal{B}$ and hence $B \in \mathcal{B}$. Hence whenever B is a basic open set containing x, then $B \cap A \neq \phi$ ($A \in \mathcal{B}$) implies $x \in \overline{A}$, for all $A \in \mathcal{B}$ implies $x \in \bigcap_{A \in \mathcal{B}} \overline{A} \neq \phi$. Now $A \subseteq \mathcal{B}$ implies $\bigcap_{A \in \mathcal{A}} \overline{A} \neq \phi$. That is, whenever $\mathcal{A}$ is a collection of closed subsets of the product space (X, J) and further $\mathcal{A}$ has f.i.p then $\bigcap_{A \in \mathcal{A}} \overline{A} \neq \phi$. This proves that (X, J) is a compact topological space.

Now let us introduce the notion of a generalized sequence, known as net and convergence of a net in a topological space.

Let $(X, \leq)$ be a partially ordered set. Further suppose for $\alpha, \beta \in X$ there exist $\gamma \in X$ such that $\alpha \le \gamma$ and $\beta \le \gamma$. Then we say that $(X, \le)$ is a directed set. (In the above case if $\alpha \le \gamma$ then we also say $\gamma \ge \alpha$.)

Definition: Let X be a nonempty set and $(D, \le)$ be a directed set. Then any function $f : D \to X$ is called a net in X. For each $\alpha \in D$, $f(\alpha) = x_\alpha \in X$ and we say that $\{x_\alpha\}_{\alpha \in D}$ is a net in X.

Example: Let $D = \mathbb{N}$ and $\le$ be the usual relation on $\mathbb{N}$. Then $(\mathbb{N}, \le)$ is a directed set. If X is a nonempty set and $f : \mathbb{N} \to X$ then for each $n \in \mathbb{N}$, $f(n) = x_n \in X$. Hence our net $\{x_n\}_{n\in\mathbb{N}}$ is the well known concept namely sequence in X. In this sense we say that every sequence is a net. Now take $D = [0, 1]$. Then $(D, \le)$ is also a directed set. Define $f : [0,1] \to \mathbb{R}$ as $f(\alpha) = \alpha + 3$, $\forall\, \alpha \in [0, 1]$. Here $f = \{f(\alpha)\}_{\alpha\in D} = \{\alpha + 3\}_{\alpha\in[0,1]}$ is a net (generalized sequence) in $\mathbb{R}$.

It is intuitively clear that the net $\{\alpha + 3\}_{\alpha\in[0,1]}$ approaches to 4. What do we mean by saying that the net $\{x_\alpha\}_{\alpha\in D}$ approaches to 4? Can we also say that the net $\{x_\alpha\}_{\alpha\in D}$ approaches 3? Well, in R consider a sequence $\{x_n\}_{n\in\mathbb{N}} = \{x_n\}_{n=1}^{\infty}$. We know that $\lim_{n\to\infty} x_n = x$ (i.e $x_n \to x$ as $n \to \infty$) if and only if for each $\epsilon > 0$ there exists $n_0 \in \mathbb{N}$ such that $x_n \in (x - \epsilon, x + \epsilon)$ for all $n \ge n_0$. Note that $x_n \to x$ as $n \to \infty$ if and only if for each open set U containing x there exists $n_0 \in \mathbb{N}$ such that $x_n \in U$ for all $n \ge n_0$.

Keeping this in mind, we define:

Let $(X, \mathcal{J})$ be a topological space and $\{x_\alpha\}_{\alpha\in D}$ be a net in X. Then we say that the net $\{x_\alpha\}_{\alpha\in D}$ converges to an element $x \in X$ if and only if for each open set U containing x there exists $\alpha_0 \in D$ such that $x_\alpha \in U$, $\forall\, \alpha \ge \alpha_0$ (that is $\alpha \in D$ with $\alpha_0 \le \alpha$). If $\{x_\alpha\}_{\alpha\in D}$ converges to x then we write $x_\alpha \to x$.

In a metric space (X, d) we know that a sequence $\{x_n\}_{n=1}^{\infty}$ in X converges to at most one element x in X.

What about in a topological space ? Whether a net $\{x_\alpha\}_{\alpha\in D}$ in a topological space converges to at most one element in X. Obviously the answer is no. For example, let X be any set containing at least two elements and $\mathcal{J} = \{\phi, X\}$. Take $x_1, x_2 \in X$, $x_1 \ne x_2$. Now with usual $\le$, $(\mathbb{N}, \le)$ is a directed set.

Define $f : \mathbb{N} \to X$ as

$$f(n) = \begin{cases} x_1 & \textit{when n is odd} \\ x_2 & \textit{when n is even} \end{cases}$$

Here our net is $\{x_1, x_2, x_1, x_2, \ldots\}$ that is our net is a sequence in X. Let $x = x_1$. Then the only open set U containing x_1 is X and hence $n_0 = 1 \in \mathbb{N}$. Then for all $n \ge n_0$, $x_n \in X = U$. Hence $x_n \to x$ for any $x \in X$. Also nothing special about the net $\{x_1, x_2, x_1, x_2, \ldots\}$. In fact if D is a directed set and $\{x_\alpha\}_{\alpha\in D}$ is an arbitrary net in X then for each $x \in X$, $x_\alpha \to x$.

Example: Now consider $X = \mathbb{R}$ and $\mathcal{J}_f$, the cofinite topology on $\mathbb{R}$. $D = \mathbb{R}$ and ≤ is our usual relation. Then (D, ≤) is a directed set. Define $f : D \to \mathbb{R}$ as f(α) = α for $\alpha \in D = \mathbb{R}$. Then $\{\alpha\}_{\alpha \in \mathbb{R}}$ is a net in $\mathbb{R}$. Fix an element say $x \in \mathbb{R}$ Whether $x_\alpha \to x$? How to start? Start with an open set U containing x in our topological space $(\mathbb{R}, \mathcal{J})$. Now $U \in \mathcal{J}_f$, $x \in U$ (that is $U \neq \phi$) implies U^c is a finite subset of $\mathbb{R}$.

Case (i). $U^c = \phi$ $(\Rightarrow U = X)$.

Case (ii). $U^c \neq \phi$

That is U^c is a nonempty finite subset of $\mathbb{R}$. Hence there exists $n_0 \in \mathbb{N}$ and $x_1, x_2, \ldots, x_{n0} \in \mathbb{R} = D$ such that $U^c = \{\alpha_1, \alpha_2, \ldots, \alpha_{n_0}\}$. Now take a real number say α_0 such that $\alpha_0 > \alpha_i$ for all i = 1, 2, . . . , n_0. This $\alpha_0 \in D$ is such that $x_\alpha = \alpha \in U \forall \alpha \geq \alpha_0, (\alpha \geq \alpha_0, \alpha_0 > \alpha_i \Rightarrow \alpha > \alpha_i \Rightarrow \alpha \notin U^c \Rightarrow \alpha \in U)$.

Conclusion: We started with an open set U containing x and we could get an $\alpha_0 \in D$ (α_0 depends on U) such that $x_\alpha \in U$, $\forall \alpha \geq \alpha_0$. Hence by our definition $x_\alpha \to x$. That this net $\{x_\alpha\} = \{\alpha\}_{\alpha \in D}$ converges to every element x of the given topological space $(\mathbb{R}, \mathcal{J})$.

(iii) D = {1, 2, . . . , p} and ≤ is our usual relation. (D, ≤) is a directed set (check).

What about $\{x_\alpha\}_{\alpha \in D}$. Here D = {1, 2, . . . , 10} implies $\{x_\alpha\}_{\alpha \in D}$ = {1, 2, . . . , 10}. Now for any open set U containing 10 there exists $\alpha_0 = 10 \in D$ is such that $\alpha \in D$, $\alpha \geq \alpha_0 = 10 \Rightarrow \alpha = 10$ and $x_\alpha = \alpha = 10 \in U$. Hence $\{x_\alpha\}_{\alpha \in D} \to 10$.

Theorem 12. In a Hausdorff topological space $(X, \mathcal{J})$ a net $\{x_\alpha\}_{\alpha \in D}$ in X cannot converge to more than one element.

Proof: Suppose a net $\{x_\alpha\}_{\alpha \in D}$ converge to say $x, y \in X$, where $x \neq y$. Now $x \neq y$, $(X, \mathcal{J})$ is a Hausdorff topological space implies there exist open sets U, V in X such that (i) $x \in U$, $y \in U$, (ii) $U \cap V = \phi$. Now $x_\alpha \to x$, U is an open set containing x implies

$$\text{there exists } \alpha_1 \in D \text{ such that } x_\alpha \in U \text{ for all } \alpha \geq \alpha_1. \qquad (5)$$

Also $y_\alpha \to y$, V is an open set containing y implies

$$\text{there exist } \alpha_2 \in D \text{ such that } y_\alpha \in V \text{ for all } \alpha \geq \alpha_2. \qquad (6)$$

Note that D with a relation ≤ is a directed set and hence for $\alpha_1, \alpha_2 \in D$ there exists $\alpha_0 \in D$ such that $\alpha_0 \geq \alpha_1$ and $\alpha_0 \geq \alpha_2$ (that is $\alpha_1 \leq \alpha_0$ and $\alpha_2 \leq \alpha_0$). Now $\alpha_0 \geq \alpha_1$ implies $x_{\alpha_0} \in U$ from Eq. (5) and $\alpha_0 \geq \alpha_2$ implies $x_{\alpha_0} \in V$ from Eq. (6). Hence $x_{\alpha_0} \in U \cap V$, a contradiction to $U \cap V = \phi$. We arrived at this contradiction by assuming $x_\alpha \to x$, $x_\alpha \to y$ and $x \neq y$. This means $\{x_\alpha\}_{\alpha \in D}$ cannot converge to more than one element.

Note. In a Hausdorff topological space a net $\{x_\alpha\}_{\alpha \in D}$ may not converge. If a net converges then it converges to a unique limit.

Theorem 13. Let $(X, \mathcal{J})$ be a topological space and $A \subseteq X$. Then an element x of X is in $\overline{A}$ if and only if there exists a net $\{x_\alpha\}_{\alpha \in D}$ in A such that $x_\alpha \to x$.

Proof: Let us assume that $x \in \overline{A}$. Our tasks are the following: (i) using the fact that $x \in \overline{A}$ construct a suitable directed set, (D, ≤), (ii) and then define a net $\{x_\alpha\}_{\alpha \in D}$ that converges to x. Now $x \in \overline{A}$ implies for each open set U containing x, $U \cap A \neq \phi$. (If our topology $\mathcal{J}$ is induced by a metric d on X then $\mathcal{J} = \mathcal{J}_d$. In this case $B\left(x, \frac{1}{n}\right) \cap A \neq \phi$ for each $n \in \mathbb{N}$. So take $x_n \in B\left(x, \frac{1}{n}\right) \cap A$. Then $d(x_n, x) < \frac{1}{n}$ and $\frac{1}{n} \to 0$ as $n \to \infty$. Hence $x_n \to x$). Take $D = \mathcal{N}_x = \{U \in \mathcal{J} : x \in U\}$ that is $\mathbb{N}_x$ is the collection of all open sets containing x. For $U, V \in \mathbb{N}_x$, define $U \leq V$ if and only if $V \subseteq U$ (reverse set inclusion is our relation ≤). Now define $f : \mathbb{N}_x \to X$ as $f(U) = x_U \in U \cap A$ ($U \cap A \neq \phi$ for each $U \in \mathbb{N}_x$ implies by axiom of choice such a function exists). Now we have a net $\{x_U\}_{U \in \mathbb{N}_x}$.

Claim: $x_U \to x$.

Take an open set U_0 containing x, then such an $U_0 \in \mathbb{N}_x$ implies $f(U_0) \in U_0 \cap A$. Now $U \in \mathbb{N}_x$ (our directed set) and $U \geq U_0$ implies $U \subseteq U_0$ implies $x_U \in U \subseteq U_0$. Now $U \geq U_0$ implies $x_U \in U_0$. Hence by definition of convergence of a net, $x_U \to x$.

Conversely, assume that there is a net say $\{x_\alpha\}_{\alpha \in D}$ in A such that $x_\alpha \to x$. Now we will have to prove that $x \in \overline{A}$. So start with an open set U containing x. Hence $x_\alpha \to x$ implies

$$\text{there exists } \alpha_0 \in D \text{ such that } x_\alpha \in U \text{ for all } \alpha \geq \alpha_0. \qquad (7)$$

(D is a directed set means (D, ≤) is a directed set). In particular when $\alpha = \alpha_0$, $\alpha \geq \alpha_0$ and therefore from Eq. (7), $x_{\alpha_0} \in U$. Also $x_{\alpha_0} \in A$. Hence $x_{\alpha_0} \in U \cap A$. That is for each open set U containing x, $U \cap A \neq \phi$. This implies $x \in \overline{A}$.

Theorem 14. Let X, Y be topological spaces and $f: X \to Y$. Then f is continuous if and only if for every net $\{x_\alpha\}_{\alpha \in J}$ converging to an element $x \in X$ the net $\{f(x_\alpha)\}_{\alpha \in J}$ converges to $f(x)$.

Proof: Assume that $f: X \to Y$ is a continuous function. Now let $\{x_\alpha\}_{\alpha \in J}$ be a net in X such that $x_\alpha \to x$ for some $x \in X$. We will have to prove that $f(x_\alpha) \to f(x)$.

Let V be an open set containing $f(x)$ in Y. Now V is an open set containing $f(x)$ and $f: X \to Y$ is a continuous function implies

$$\text{there exists an open set U containing } x \text{ such that } f(U) \subseteq V. \qquad (8)$$

Now U is an open set containing x and $x_\alpha \to x$ implies there exists an $\alpha_0 \in J$ such that $x_\alpha \in U$ for all $\alpha \geq \alpha_0$. Hence from Eq. (4.8), $f(x_\alpha) \in V$ for all $\alpha \geq \alpha_0$. That is, for each open set V containing $f(x)$ there exists $\alpha_0 \in J$ such that $f(x_\alpha) \in V$ for all $\alpha \geq \alpha_0$. This in turn implies $f(x_\alpha) \to f(x)$.

Now let us assume that whenever a net $\{x_\alpha\}_{\alpha\in J}$ converges to an element x in X then $f(x_\alpha) \to f(x)$ in Y . In this case we will have to prove that $f : X \to Y$ continuous. We know that f is continuous if and only if $f(\overline{A}) \subseteq \overline{f(A)}$ for all $A \subseteq X$. (An element z of X is closer to A, that is if $z \in \overline{A}$ then the image f(z) is closer to f(A).) So start with $A \subseteq X$ and an element $y \in f(\overline{A})$ $(f(\overline{A}) = \phi \Rightarrow f(\overline{A}) \subseteq \overline{f(A)})$. Now $y \in f(\overline{A})$ implies there exists $x \in \overline{A}$ such that $y = f(x)$. Hence $x \in \overline{A}$ implies there exists a net $\{x_\alpha\}_{\alpha\in J}$ in A such that $x_\alpha \to x$ (refer the previous theorem) this implies by our assumption, $f(x_\alpha) \to f(x)$. Now $f(x_\alpha)$ $\in f(A)$ and $f(x_\alpha) \to f(x)$ implies $f(x) \in \overline{f(A)}$ (again refer the previous theorem). So we have proved that $f(\overline{A}) \subseteq \overline{f(A)}$ whenever $A \subseteq X$. This implies $f : X \to Y$ is a continuous function.

Alternate Proof of Theorem

Proof: Assume that whenever a net $\{x_\alpha\}_{\alpha\in J}$ converges to an element $x \in X$ then $f(x_\alpha) \to f(x)$ in Y. Now suppose f is not continuous at x. Then there exists an open set V containing x such that $f(U) \not\subseteq V$ for every $U \in \mathcal{N}_x$. Then for each $U \in \mathcal{N}_x$ there exists $x_U \in U$ such that $f(x_U) \notin V$. Now observe (refer the proof of the theorem 13) that the net $\{x_U\}_{U\in\mathcal{N}_x}$ such that $x_U \to x$ in X but $f(x_U) \not\to f(x)$ in Y.

Now let us define a concept which generalize the concept of a subsequence. Recall that if X is a nonempty set $\{x_n\}_{n=1}^{\infty} = (x_n)_{n\in\mathbb{N}}$ is a sequence in X if and only if there exists a function $f : \mathbb{N} \to X$ satisfying the condition that $f(x_n) = x_n$. Here we have a net $\{x_\alpha\}_{\alpha\in J}$ in X. Hence in place of $\mathbb{N}$ we have a directed set $(J, \leq)$ and a function $f : J \to X$ satisfying the condition that $f(\alpha) = x_\alpha$ for all $\alpha \in J$.

What do we mean by saying that $\{x_{n_k}\}_{k=1}^{\infty}$ is a subsequence of $\{x_n\}$? We have a subset $\{n_k\}_{k=1}^{\infty}$ of natural numbers satisfying $n_1 < n_2 < \cdots < n_k < n_{k+1} < \cdots$. Let $D = \{n_k : k \in \mathbb{N}\}$ and we have $f : \mathbb{N} \to X$ such that $f(n) = x_n$ for all $n \in \mathbb{N}$. That is essentially $f : \mathbb{N} \to X$ is sequence in X. Now we have another function say $g : D \to \mathbb{N}$ satisfying (i) $g(k) = n_k$, (ii) $n_1 < n_2 < \cdots < n_k < n_{k+1} < \cdots$ that is $k < l \Rightarrow n_k < n_l$ that is $k < l \Rightarrow g(k) < g(l)$. Also note that for each $n_0 \in \mathbb{N}$ there exists $k_0 \in \mathbb{N}$ such that $f(k_0) = n_{k_0} > n_0$. So keeping this motivation in mind we define the concept of subnet of a given net in X. It is given that $f : J \to X$ is a net in X (so it is understood that $(J, \leq)$ is a directed set). Suppose D with a relation $\leq$ is a directed set (need not be the same relation as given in J. But for the sake of simplicity we use same notation $\leq$ for both sets). Suppose $g : D \to J$ such that $i, j \in D$, $i \leq j$ implies $g(i) \leq g(j)$ and for each $\alpha \in J$ there exists $\gamma \in D$ such that $g(\gamma) \geq \alpha$ (this is like saying that, when $J = \mathbb{N} = D$, for each $n_0 \in \mathbb{N}$ there exists $k \in \mathbb{N}$ such that $g(k) = n_k \geq n_0$). In such a case $f \circ g : D \to X$ is called subnet of X. ($(f \circ g)(k) = f(g(k)) = f(n_k) = x_{n_k}$).

Definition: Let $(X, \mathcal{J})$ be a topological space and $\{x_\alpha\}_{\alpha\in J} = (x_\alpha)_{\alpha\in J}$ be a net in X. An element $x \in X$ is said to be an accumulation point of the given net $(x_\alpha)_{\alpha\in J}$ if and only if for each open set U containing x, the set $K_U = \{\alpha \in J : x_\alpha \in U\}$ is cofinal in J. Now K_U is cofinal in J means for each $\alpha \in J$ there exists $\beta \in K_U$ such that $\beta \geq \alpha$ (it is like saying that $k \to \infty$ implies $n_k \to \infty$). Now let us prove:

Theorem 15. Let $(x_\alpha)_{\alpha \in J}$ be a net in a topological space. Then a point x in X is an accumulation point of the given net $(x_\alpha)_{\alpha \in J}$ if and only if $(x_\alpha)_{\alpha \in J}$ has a subnet and that subnet converges to x.

Proof: ⇒ Assume that x is an accumulation point of $(x_\alpha)_{\alpha \in J}$. By the definition of accumulation point of a net we have for each open set U containing x

$$K_U = \{\alpha \in J : x_\alpha \in U\} \text{ is cofinal in } J. \qquad (9)$$

Let $K = \{(\alpha, U) \in J \times \mathcal{N}_x : x_\alpha \in U\}$, where $\mathcal{N}_x$ is the collection of all open sets containing x. From Eq. (9), $K_U \neq \phi$. (Fix $\alpha \in J$. Now K_U is cofinal in J implies there exists $\beta \in K_U$ such that $\beta \geq \alpha$.) For $(\alpha, U),(\beta, V) \in K$ define $(\alpha, U) \leq (\beta, V)$ if and only if $\alpha \leq \beta$ and $V \subseteq U$ (reverse set inclusion). It is easy to see that $(K, \leq)$ is a directed set. It is given that $(x_\alpha)_{\alpha \in J}$ is a net in X. Hence $(J, \leq)$ is a directed set and $f : J \to X$ is such that $f(\alpha) = x_\alpha$. Now define $g : K \to J$ as $g(\alpha, U) = \alpha$ (refer Eq. (9)).

Claim: $g(K)$ is cofinal in J.

So take $\alpha \in J$. Now K_U is cofinal in J (refer Eq. (9)) there exists $\beta \in K_U$ such that $\beta \geq \alpha$. Now $\beta \in K_U$ implies $x_\beta \in U$ that is $(\beta, U) \in K$ is such that $g(\beta, U) = \beta \geq \alpha$ implies $g(K)$ is cofinal in J. Also $(\alpha, U),(\beta, V) \in K,(\alpha, U) \leq (\beta, V)$ implies $g(\alpha, U) = \alpha \leq \beta = g(\beta, V)$. Hence $f \circ g : K \to X$ is a subnet of f (or say $f(\alpha) = (x_\alpha)$). Now let us prove that this subnet converges to x. So take an open set U containing x. This implies K_U is cofinal in J. Fix $(\alpha_0, U) \in K$. Now $\alpha_0 \in J$, K_U is cofinal in J implies $\beta_0 \in K_U$ such that $\beta_0 \geq \alpha_0$. Hence $(\alpha, V) \in K$, $(\alpha, V) \geq (\alpha_0, U)$ implies $(f \circ g)(\alpha, V) = f(\alpha) = x_\alpha \in V \subseteq U$. That is for each open set U containing x there exists $(\alpha_0, U) \in K$ such that $(\alpha, V) \in K$ $(\alpha, V) \geq (\alpha_0, U)$ implies $(f \circ g)(\alpha, V) \in U$. This proves that $f \circ g \to x$.

Conversely, suppose there is a subnet of $(f(\alpha))_{\alpha \in J} = (x_\alpha)_{\alpha \in J}$ which converge to an element $x \in X$. A subnet of f converges to x means there exists a directed set say $(K, \leq)$ and a function say $g : K \to J$ such that $i, j \in K$, $i \leq j$ implies $g(i) \leq g(j)$, $g(K)$ is cofinal in J, and $(f \circ g)(i) = f(g(i)) \to x$. Now let us prove that x is an accumulation point of the net f. So take an open set U containing x.

Claim: $\{\alpha \in J : f(\alpha) = x_\alpha \in U\}$ is cofinal in J.

Let $\alpha_0 \in J$. Now $f \circ g : K \to X$ is a subnet such that $f \circ g \to x$. Hence for this given $\alpha_0 \in J$ there exists $\beta \in K$ such that $g(\beta) \geq \alpha_0$ (note $g(K)$ is cofinal in J). Now $f \circ g \to x$, U is an open set containing x implies there exists $\beta_0 \in K$ such that $\alpha \in K$, $\alpha \geq \beta_0 \Rightarrow f(g(\alpha)) \in U$, $\beta \in J$ is such that $g(\beta) \geq \alpha_0$. Take $\gamma_0 \in K$ such that $\alpha_0 \geq \beta, \beta_0$. Then $(f \circ g)(\gamma_0) \in U$ and $g(\gamma_0) \geq g(\beta) \geq \alpha_0$. That is for $\alpha_0 \in J$, there exists $g(\gamma_0) \in J$ such that $f(g(\gamma_0)) \in U$ implies $\{\alpha \in J : f(\alpha) = x_\alpha \in U\}$ is cofinal in J. Hence x is an accumulation point.

Recall that a metric space (X, d) is a compact metric space if and only if every sequence $\{x_n\}_{n=1}^{\infty}$ in X has a subsequence $\{x_{n_k}\}_{k=1}^{\infty}$ that converges to an element in X. It is to be noted that this result is not true for an arbitrary topological space. For a topological space we have the following theorem.

Theorem 16. A topological space $(X, \mathcal{J})$ is compact if and only if every net in X has a subnet that converges to an element in X.

Proof: Assume that $(X, \mathcal{J})$ is a compact topological space and $f : J \to X$ is a net in X. We will have to prove that f has a subnet that converges to an element in X. So it is enough to prove that f has an accumulation point.

For each $\alpha \in J$, let $A_\alpha = \{x_\beta : \alpha \le \beta\}$ (note: $f : J \to X$ is a net means with respect to a relation $\le$, $(J, \le)$ is directed set). Now $\{A_\alpha\}_{\alpha \in J}$ is a collection of sets which has finite intersection property. For $A_{\alpha_1}, A_{\alpha_2}, \ldots, A_\alpha$ if we take $\alpha \in J$ such that $\alpha \ge \alpha_j$ for all $j = 1, 2, \ldots, k$, that is $\alpha_j \le \alpha$, then $x_\alpha \in A_{\alpha_j}, \forall j = 1, 2, \ldots, k$ and hence $x \in \bigcap_{j=1}^{k} A_{\alpha_j}$. Now $(X, \mathcal{J})$ is a compact topological space $\{\overline{A}_\alpha\}_{\alpha \in J}$ is a collection of closed subsets of X which has finite intersection property implies $\bigcap_{\alpha \in J} \overline{A}_{\alpha_j} \ne \phi$. Let $x \in \bigcap_{\alpha \in J} \overline{A}_{\alpha_j}$.

Now we aim to prove that x is an accumulation point of f. So, start with an open set U containing x, and we will have to prove that $\{\alpha \in J : x_\alpha \in U\}$ is cofinal in J. Take $\alpha_0 \in J$. Now U is an open set containing x, $x \in \overline{A}_{\alpha_0}$ implies $U \cap A_{\alpha_0} \ne \phi$. Hence there exists $\alpha \ge \alpha_0$ such that $x_\alpha \in U$. This proves that $\{\alpha \in J : x_\alpha \in U\}$ is cofinal in J. Hence we have proved that x is an accumulation point of the stated net f. This implies there exists a subnet of f which converges to f.

To prove the converse part let us assume that every net in X has convergent subnet in X. By assuming this, we aim to prove that $(X, \mathcal{J})$ is a compact topological space.

To prove that $(X, \mathcal{J})$ is a compact topological space, let us prove: if $\mathcal{A}$ is a collection of closed subsets of X which has finite intersection property then $\bigcap_{A \in \mathcal{A}} A \ne \phi$. So, we have a collection $\mathcal{A}$ of closed subsets of X which has finite intersection property.

Let $\mathcal{B} = \{A \subseteq X : A = A_1 \cap A_2 \cap \cdots \cap A_k, k \in \mathbb{N}, A_1, \ldots, A_k \in \mathcal{A}\}$. That is $\mathcal{B}$ is the collection of finite intersection of members of $\mathcal{A}$. (Note. $\bigcap_{A \in \phi} A = X$ and hence we do not require to consider this case.) For A, B $\in \mathcal{B}$ define $A \le B$, whenever $B \subseteq A$. Then $(\mathcal{B}, \le)$ is a directed set. Now define f: $\mathcal{B} \to X$ as $f(A) = f(A_1 \cap A_2 \cap \cdots \cap A_k) = x_A$, where $x_A \in A_1 \cap A_2 \cap \cdots \cap A_k$ is fixed ($A_1 \cap A_2 \cap \cdots \cap A_k$) may contains more than one element and in that case first take any one element form $A_1 \cap A_2 \cap \cdots \cap A_k$. Hence $f = (f(A))_{A \in \mathcal{B}}$ is a net in X. By our assumption this net f will have a subnet that will converge to an element say x in X. So there will exists a directed set K and a function $g : K \to B$ satisfying $f \circ g$ is a subnet of f and $f \circ g$ converges to x.

Now we claim that $x \in A$ for each $A \in \mathcal{A}$. Suppose for some $A \in \mathcal{A}$, $x \notin A$. Then $x \in A^c = U$, an open set. Since $f \circ g \to x$ and U is an open set containing x there exists $\alpha_0 \in K$ such that $(f \circ g)(\alpha) \in U$ for all $\alpha \geq \alpha_0$. Now $\alpha_0 \in K$ implies $g(\alpha_0) \in B$ implies there exists $k \in \mathbb{N}$ and $A_1, A_2, \ldots, A_k \in \mathcal{A}$ such that $g(\alpha_0) = A_1 \cap A_2 \cap \cdots \cap A_k$. $A_1 \cap A_2 \cap \cdots \cap A_k \in \mathcal{B}$ is such that $A_1 \cap A_2 \cap \cdots \cap A_k \geq g(\alpha_0)$. We have $f \circ g(\alpha_0) = f(g(\alpha_0)) \in U = A^c$. Now K is a directed set and g(K) is cofinal in $\mathcal{B}$ implies there exists $\alpha \in K$ such that (i) $\alpha \geq \alpha_0$, (ii) $g(\alpha) \geq A_1 \cap A_2 \cap \cdots \cap A_k$. Now $\alpha \geq \alpha_0$ implies $(f \circ g)(\alpha) \in U = A^c$ but by the definition of f, $f(g(\alpha)) \in g(\alpha) \subseteq A_1 \cap A_2 \cap \cdots \cap A_k \subseteq A$. So we get a contradiction. Therefore $x \notin A$ for some $A \in \mathcal{A}$ cannot happen. We have proved that if $\mathcal{A}$ is a collection of closed subsets of X which has finite intersection property then $\bigcap_{A \in \mathcal{A}} A \neq \phi$. Hence $(X, \mathcal{J})$ is a compact topological space.

Definition: A topological property is any property so that if $(X, \mathcal{J})$, $(Y, \mathcal{J}')$ are topological spaces and $f:(X, \mathcal{J}) \to (Y, \mathcal{J}')$ is a homeomorphism (that is $(X, \mathcal{J})$ is homeomorphic to $(Y, \mathcal{J}')$) then $(X, \mathcal{J})$ has the property if and only if $(Y, \mathcal{J}')$ has the same property.

Example: Compactness, connectedness, local compactness are all topological properties.

Compactness Theorem

In mathematical logic, the compactness theorem states that a set of first-order sentences has a model if and only if every finite subset of it has a model. This theorem is an important tool in model theory, as it provides a useful method for constructing models of any set of sentences that is finitely consistent.

The compactness theorem for the propositional calculus is a consequence of Tychonoff's theorem (which says that the product of compact spaces is compact) applied to compact Stone spaces; hence, the theorem's name. Likewise, it is analogous to the finite intersection property characterization of compactness in topological spaces: a collection of closed sets in a compact space has a non-empty intersection if every finite subcollection has a non-empty intersection.

The compactness theorem is one of the two key properties, along with the downward Löwenheim–Skolem theorem, that is used in Lindström's theorem to characterize first-order logic. Although there are some generalizations of the compactness theorem to non-first-order logics, the compactness theorem itself does not hold in them.

History

Kurt Gödel proved the countable compactness theorem in 1930. Anatoly Maltsev proved the uncountable case in 1936.

Applications

The compactness theorem has many applications in model theory; a few typical results are sketched here.

The compactness theorem implies Robinson's principle: If a first-order sentence holds in every field of characteristic zero, then there exists a constant p such that the sentence holds for every field of characteristic larger than p. This can be seen as follows: suppose φ is a sentence that holds in every field of characteristic zero. Then its negation $\neg\varphi$, together with the field axioms and the infinite sequence of sentences $1+1 \neq 0$, $1+1+1 \neq 0$, ..., is not satisfiable (because there is no field of characteristic 0 in which $\neg\varphi$ holds, and the infinite sequence of sentences ensures any model would be a field of characteristic 0). Therefore, there is a finite subset A of these sentences that is not satisfiable. We can assume that A contains $\neg\varphi$, the field axioms, and, for some k, the first k sentences of the form $1+1+...+1 \neq 0$ (because adding more sentences doesn't change unsatisfiability). Let B contain all the sentences of A except $\neg\varphi$. Then any field with a characteristic greater than k is a model of B, and $\neg\varphi$ together with B is not satisfiable. This means that φ must hold in every model of B, which means precisely that φ holds in every field of characteristic greater than k.

A second application of the compactness theorem shows that any theory that has arbitrarily large finite models, or a single infinite model, has models of arbitrary large cardinality (this is the Upward Löwenheim–Skolem theorem). So, for instance, there are nonstandard models of Peano arithmetic with uncountably many 'natural numbers'. To achieve this, let T be the initial theory and let κ be any cardinal number. Add to the language of T one constant symbol for every element of κ. Then add to T a collection of sentences that say that the objects denoted by any two distinct constant symbols from the new collection are distinct (this is a collection of κ^2 sentences). Since every *finite* subset of this new theory is satisfiable by a sufficiently large finite model of T, or by any infinite model, the entire extended theory is satisfiable. But any model of the extended theory has cardinality at least κ.

A third application of the compactness theorem is the construction of nonstandard models of the real numbers, that is, consistent extensions of the theory of the real numbers that contain "infinitesimal" numbers. To see this, let Σ be a first-order axiomatization of the theory of the real numbers. Consider the theory obtained by adding a new constant symbol ε to the language and adjoining to Σ the axiom $\varepsilon > 0$ and the axioms $\varepsilon < 1/n$ for all positive integers n. Clearly, the standard real numbers R are a model for every finite subset of these axioms, because the real numbers satisfy everything in Σ and, by suitable choice of ε, can be made to satisfy any finite subset of the axioms about ε. By the compactness theorem, there is a model *R that satisfies Σ and also contains an infinitesimal element ε. A similar argument, adjoining axioms $\omega > 0$, $\omega > 1$, etc., shows that the existence of infinitely large integers cannot be ruled out by any axiomatization Σ of the reals.

Proofs

One can prove the compactness theorem using Gödel's completeness theorem, which establishes that a set of sentences is satisfiable if and only if no contradiction can be proven from it. Since proofs are always finite and therefore involve only finitely many of the given sentences, the compactness theorem follows. In fact, the compactness theorem is equivalent to Gödel's completeness theorem, and both are equivalent to the Boolean prime ideal theorem, a weak form of the axiom of choice.

Gödel originally proved the compactness theorem in just this way, but later some "purely semantic" proofs of the compactness theorem were found, i.e., proofs that refer to *truth* but not to *provability*. One of those proofs relies on ultraproducts hinging on the axiom of choice as follows:

Proof: Fix a first-order language L, and let Σ be a collection of L-sentences such that every finite subcollection of L-sentences, $i \subseteq \Sigma$ of it has a model $\mathcal{M}_i$. Also let $\prod_{i \subseteq \Sigma} \mathcal{M}_i$ be the direct product of the structures and I be the collection of finite subsets of Σ. For each i in I let $A_i := \{j \in I : j \supseteq i\}$. The family of all of these sets A_i generates a proper filter, so there is an ultrafilter U containing all sets of the form A_i.

Now for any formula φ in Σ we have:

- the set $A_{\{\varphi\}}$ is in U
- whenever $j \in A_{\{\varphi\}}$, then $\varphi \in j$, hence φ holds in $\mathcal{M}_j$
- the set of all j with the property that φ holds in $\mathcal{M}_j$ is a superset of $A_{\{\varphi\}}$, hence also in UUsing Łoś›s theorem we see that φ holds in the ultraproduct $\prod_{i \subseteq \Sigma} \mathcal{M}_i / U$. So this ultraproduct satisfies all formulas in Σ.

Closed Set

In geometry, topology, and related branches of mathematics, a closed set is a set whose complement is an open set. In a topological space, a closed set can be defined as a set which contains all its limit points. In a complete metric space, a closed set is a set which is closed under the limit operation.

Equivalent Definitions of a Closed Set

In a topological space, a set is closed if and only if it coincides with its closure. Equivalently, a set is closed if and only if it contains all of its limit points.

Properties of Closed Sets

A closed set contains its own boundary. In other words, if you are "outside" a closed set, you may move a small amount in any direction and still stay outside the set. Note that this is also true if the boundary is the empty set, e.g. in the metric space of rational numbers, for the set of numbers of which the square is less than 2.

- Any intersection of closed sets is closed (including intersections of infinitely many closed sets)
- The union of *finitely many* closed sets is closed.
- The empty set is closed.
- The whole set is closed.

In fact, given a set *X* and a collection F of subsets of *X* that has these properties, then F will be the collection of closed sets for a unique topology on *X*. The intersection property also allows one to define the closure of a set *A* in a space *X*, which is defined as the smallest closed subset of *X* that is a superset of *A*. Specifically, the closure of *A* can be constructed as the intersection of all of these closed supersets.

Sets that can be constructed as the union of countably many closed sets are denoted F_σ sets. These sets need not be closed.

Examples of Closed Sets

- The closed interval [*a*,*b*] of real numbers is closed.
- The unit interval [0,1] is closed in the metric space of real numbers, and the set [0,1] ∩ Q of rational numbers between 0 and 1 (inclusive) is closed in the space of rational numbers, but [0,1] ∩ Q is not closed in the real numbers.
- Some sets are neither open nor closed, for instance the half-open interval [0,1) in the real numbers.
- Some sets are both open and closed and are called clopen sets.
- The ray [1, +∞) is closed.
- The Cantor set is an unusual closed set in the sense that it consists entirely of boundary points and is nowhere dense.
- Singleton points (and thus finite sets) are closed in Hausdorff spaces.
- The set of integers Z is an infinite and unbounded closed set in the real numbers.
- If *X* and *Y* are topological spaces, a function *f* from *X* into *Y* is continuous if and only if preimages of closed sets in *Y* are closed in *X*.

More About Closed Sets

In point set topology, a set *A* is closed if it contains all its boundary points.

The notion of closed set is defined above in terms of open sets, a concept that makes sense for topological spaces, as well as for other spaces that carry topological structures, such as metric spaces, differentiable manifolds, uniform spaces, and gauge spaces.

An alternative characterization of closed sets is available via sequences and nets. A subset A of a topological space X is closed in X if and only if every limit of every net of elements of A also belongs to A. In a first-countable space (such as a metric space), it is enough to consider only convergent sequences, instead of all nets. One value of this characterization is that it may be used as a definition in the context of convergence spaces, which are more general than topological spaces. Notice that this characterization also depends on the surrounding space X, because whether or not a sequence or net converges in X depends on what points are present in X.

Whether a set is closed depends on the space in which it is embedded. However, the compact Hausdorff spaces are "absolutely closed", in the sense that, if you embed a compact Hausdorff space K in an arbitrary Hausdorff space X, then K will always be a closed subset of X; the "surrounding space" does not matter here. Stone-Čech compactification, a process that turns a completely regular Hausdorff space into a compact Hausdorff space, may be described as adjoining limits of certain nonconvergent nets to the space.

Furthermore, every closed subset of a compact space is compact, and every compact subspace of a Hausdorff space is closed.

Closed sets also give a useful characterization of compactness: a topological space X is compact if and only if every collection of nonempty closed subsets of X with empty intersection admits a finite subcollection with empty intersection.

A topological space X is disconnected if there exist disjoint, nonempty, closed subsets A and B of X whose union is X. Furthermore, X is totally disconnected if it has an open basis consisting of closed sets.

Euclidean Space

In geometry, Euclidean space encompasses the two-dimensional Euclidean plane, the three-dimensional space of Euclidean geometry, and certain other spaces. It is named after the Ancient Greek mathematician Euclid of Alexandria. The term "Euclidean" distinguishes these spaces from other types of spaces considered in modern geometry. Euclidean spaces also generalize to higher dimensions.

Classical Greek geometry defined the Euclidean plane and Euclidean three-dimensional space using certain postulates, while the other properties of these spaces were deduced as theorems. Geometric constructions are also used to define rational numbers.

When algebra and mathematical analysis became developed enough, this relation reversed and now it is more common to define Euclidean space using Cartesian coordinates and the ideas of analytic geometry. It means that points of the space are specified with collections of real numbers, and geometric shapes are defined as equations and inequalities. This approach brings the tools of algebra and calculus to bear on questions of geometry and has the advantage that it generalizes easily to Euclidean spaces of more than three dimensions.

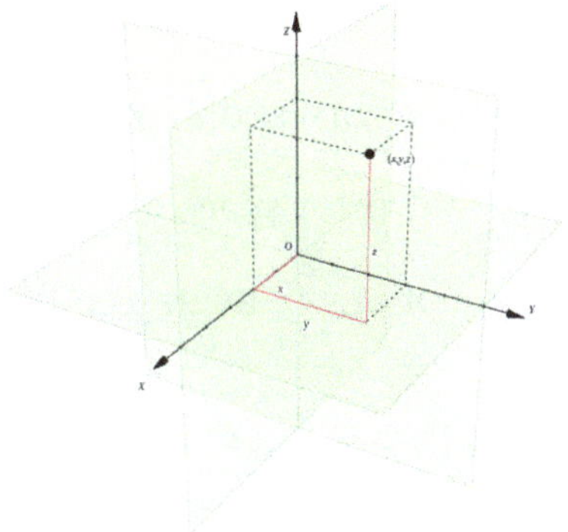

Every point in three-dimensional Euclidean space is determined by three coordinates.

From the modern viewpoint, there is essentially only one Euclidean space of each dimension. With Cartesian coordinates it is modelled by the real coordinate space (R^n) of the same dimension. In one dimension, this is the real line; in two dimensions, it is the Cartesian plane; and in higher dimensions it is a coordinate space with three or more real number coordinates. Mathematicians denote the n-dimensional Euclidean space by E^n if they wish to emphasize its Euclidean nature, but R^n is used as well since the latter is assumed to have the standard Euclidean structure, and these two structures are not always distinguished. Euclidean spaces have finite dimension.

Intuitive Overview

One way to think of the Euclidean plane is as a set of points satisfying certain relationships, expressible in terms of distance and angle. For example, there are two fundamental operations (referred to as motions) on the plane. One is translation, which means a shifting of the plane so that every point is shifted in the same direction and by the same distance. The other is rotation about a fixed point in the plane, in which every point in the plane turns about that fixed point through the same angle. One of the basic tenets of Euclidean geometry is that two figures (usually considered as subsets) of the plane should be considered equivalent (congruent) if one can be transformed into the other by some sequence of translations, rotations and reflections.

In order to make all of this mathematically precise, the theory must clearly define the notions of distance, angle, translation, and rotation for a mathematically described space. Even when used in physical theories, Euclidean space is an abstraction detached from actual physical locations, specific reference frames, measurement instruments, and so on. A purely mathematical definition of Euclidean space also ignores questions of units of length and other physical dimensions: the distance in a "mathematical" space is a number, not something expressed in inches or metres. The standard way to define such space, as carried out in the remainder of this article, is to define the Euclidean plane as a two-dimensional real vector space equipped with an inner product. The reason for working with arbitrary vector spaces instead of R^n is that it is often preferable to work in a *coordinate-free* manner (that is, without choosing a preferred basis). For then:

- the vectors in the vector space correspond to the points of the Euclidean plane,
- the addition operation in the vector space corresponds to translation, and
- the inner product implies notions of angle and distance, which can be used to define rotation.

Once the Euclidean plane has been described in this language, it is actually a simple matter to extend its concept to arbitrary dimensions. For the most part, the vocabulary, formulae, and calculations are not made any more difficult by the presence of more dimensions. (However, rotations are more subtle in high dimensions, and visualizing high-dimensional spaces remains difficult, even for experienced mathematicians.)

A Euclidean space is not technically a vector space but rather an affine space, on which a vector space acts by translations, or, conversely, a Euclidean vector is the difference (displacement) in an ordered pair of points, not a single point. Intuitively, the distinction says merely that there is no canonical choice of where the origin should go in the space, because it can be translated anywhere. When a certain point is chosen, it can be declared the origin and subsequent calculations may ignore the difference between a point and its coordinate vector, as said above.

Euclidean Structure

These are distances between points and the angles between lines or vectors, which satisfy certain conditions, which makes a set of points a Euclidean space. The natural way to obtain these quantities is by introducing and using the standard inner product (also known as the dot product) on R^n. The inner product of any two real n-vectors x and y is defined by

$$\mathrm{x} \cdot \mathrm{y} = \sum_{i=1}^{n} x_i y_i = x_1 y_1 + x_2 y_2 + \cdots + x_n y_n,$$

where x_i and y_i are ith coordinates of vectors x and y respectively. The result is always a real number.

Distance

The inner product of x with itself is always non-negative. This product allows us to define the "length" of a vector x through square root:

$$\| \mathrm{x} \| = \sqrt{\mathrm{x} \cdot \mathrm{x}} = \sqrt{\sum_{i=1}^{n} (x_i)^2}.$$

This length function satisfies the required properties of a norm and is called the Euclidean norm on R^n.

Finally, one can use the norm to define a metric (or distance function) on R^n by

$$d(\mathrm{x}, \mathrm{y}) = \| \mathrm{x} - \mathrm{y} \| = \sqrt{\sum_{i=1}^{n} (x_i - y_i)^2}.$$

This distance function is called the Euclidean metric. This formula expresses a special case of the Pythagorean theorem.

This distance function (which makes a metric space) is sufficient to define all Euclidean geometry, including the dot product. Thus, a real coordinate space together with this Euclidean structure is called Euclidean space. Its vectors form an inner product space (in fact a Hilbert space), and a normed vector space.

The metric space structure is the main reason behind the use of real numbers R, not some other ordered field, as the mathematical foundation of Euclidean (and many other) spaces. Euclidean space is a complete metric space, a property which is impossible to achieve operating over rational numbers, for example.

Angle

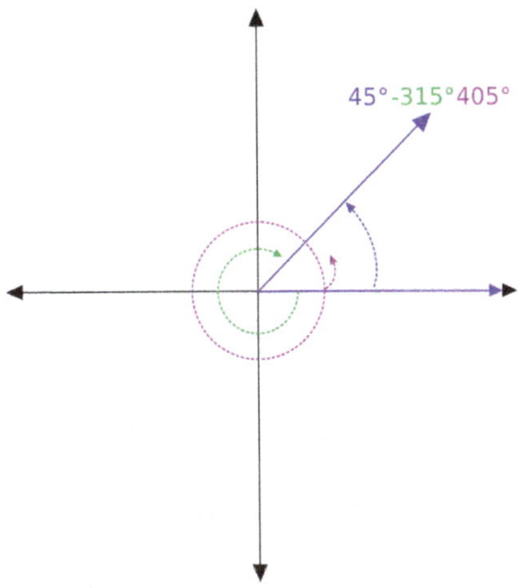

Positive and negative angles on the oriented plane

The (non-reflex) angle θ ($0° \leq \theta \leq 180°$) between vectors x and y is then given by

$$\theta = \arccos\left(\frac{\mathrm{x} \cdot y}{\|\mathrm{x}\| \; \|y\|}\right)$$

where arccos is the arccosine function. It is useful only for $n > 1$, and the case $n = 2$ is somewhat special. Namely, on an oriented Euclidean plane one can define an angle between two vectors as a number defined modulo 1 turn (usually denoted as either 2π or $360°$), such that $\angle \mathrm{yx} = -\angle \mathrm{xy}$. This oriented angle is equal either to the angle θ from the formula above or to $-\theta$. If one non-zero vector is fixed (such as the first basis vector), then each non-zero vector is uniquely defined by its magnitude and angle.

The angle does not change if vectors x and y are multiplied by positive numbers.

Unlike the aforementioned situation with distance, the scale of angles is the same in pure mathematics, physics, and computing. It does not depend on the scale of distances; all distances may be multiplied by some fixed factor, and all angles will be preserved. Usually, the angle is considered a dimensionless quantity, but there are different units of measurement, such as radian (preferred in pure mathematics and theoretical physics) and degree (°) (preferred in most applications).

Rotations and Reflections

Symmetries of a Euclidean space are transformations which preserve the Euclidean metric (called *isometries*). Although aforementioned translations are most obvious of them, they have the same structure for any affine space and do not show a distinctive character of Euclidean geometry. Another family of symmetries leave one point fixed, which may be seen as the origin without loss of generality. All transformations, which preserves the origin and the Euclidean metric, are linear maps. Such transformations Q must, for any x and y, satisfy:

$Q\mathrm{x} \cdot Q\mathrm{y} = \mathrm{x} \cdot \mathrm{y}$ (explain the notation),

$|Q\mathrm{x}| = |\mathrm{x}|.$

Such transforms constitute a group called the *orthogonal group* $\mathrm{O}(n)$. Its elements Q are exactly solutions of a matrix equation

$$Q^{\mathrm{T}} Q = Q Q^{\mathrm{T}} = I,$$

where Q^{T} is the transpose of Q and I is the identity matrix.

But a Euclidean space is orientable. Each of these transformations either preserves or reverses orientation depending on whether its determinant is +1 or −1 respectively. Only transformations which preserve orientation, which form the *special orthogonal* group SO(n), are considered (proper) rotations. This group has, as a Lie group, the same dimension $n(n - 1)/2$ and is the identity component of O(n).

Group	Diffeo-morphic to	Isomorphic to
SO(1)	{1}	
SO(2)	S^1	U(1)
SO(3)	$\mathbf{RP}^3$	SU(2)/{±1}
SO(4)	$(S^3 \times S^3)/\{\pm 1\}$	(SU(2) × SU(2))/{±1}
Note: elements of SU(2) are also known as versors.		

Groups SO(n) are well-studied for $n \leq 4$. There are no non-trivial rotations in 0- and 1-spaces. Rotations of a Euclidean plane ($n = 2$) are parametrized by the angle (modulo 1 turn). Rotations of a 3-space are parametrized with axis and angle, whereas a rotation of a 4-space is a superposition of two 2-dimensional rotations around perpendicular planes.

Among linear transforms in O(n) which reverse the orientation are hyperplane reflections. This is the only possible case for $n \leq 2$, but starting from three dimensions, such isometry in the general position is a rotoreflection.

Euclidean Group

The Euclidean group $E(n)$, also referred to as the group of all isometries ISO(n), treats translations, rotations, and reflections in a uniform way, considering them as group actions in the context of group theory, and especially in Lie group theory. These group actions preserve the Euclidean structure.

As the group of all isometries, ISO(n), the Euclidean group is important because it makes Euclidean geometry a case of Klein geometry, a theoretical framework including many alternative geometries.

The structure of Euclidean spaces – distances, lines, vectors, angles (up to sign), and so on – is invariant under the transformations of their associated Euclidean group. For instance, translations form a commutative subgroup that acts freely and transitively on E^n, while the stabilizer of any point there is the aforementioned O(n).

Along with translations, rotations, reflections, as well as the identity transformation, Euclidean motions comprise also glide reflections (for $n \geq 2$), screw operations and rotoreflections (for $n \geq 3$), and even more complex combinations of primitive transformations for $n \geq 4$.

The group structure determines which conditions a metric space needs to satisfy to be a Euclidean space:

1. Firstly, a metric space must be translationally invariant with respect to some (finite-dimensional) real vector space. This means that the space itself is an affine space, that the space is *flat*, not curved, and points do not have different properties, and so any point can be translated to any other point.
2. Secondly, the metric must correspond in the aforementioned way to some positive-defined quadratic form on this vector space, because point stabilizers have to be isomorphic to O(n).

Non-Cartesian Coordinates

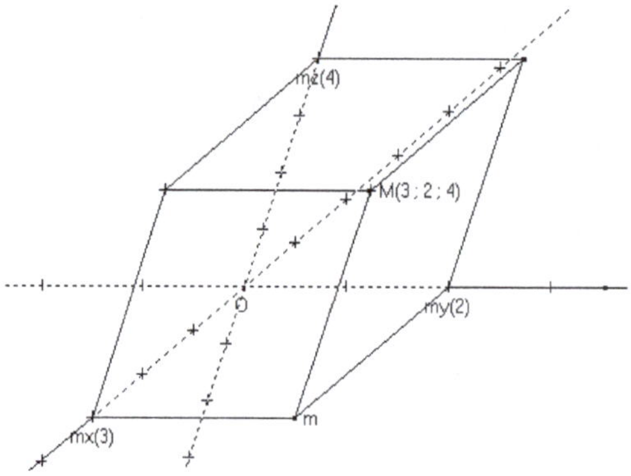

3-dimensional skew coordinates

Cartesian coordinates are arguably the standard, but not the only possible option for a Euclidean space. Skew coordinates are compatible with the affine structure of E^n, but make formulae for angles and distances more complicated.

Another approach, which goes in line with ideas of differential geometry and conformal geometry, is orthogonal coordinates, where coordinate hypersurfaces of different coordinates are orthogonal, although curved. Examples include the polar coordinate system on Euclidean plane, the second important plane coordinate system.

Geometric Shapes

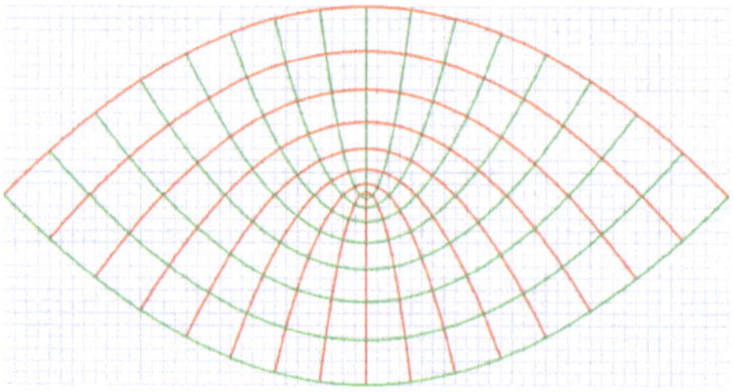

Parabolic coordinates

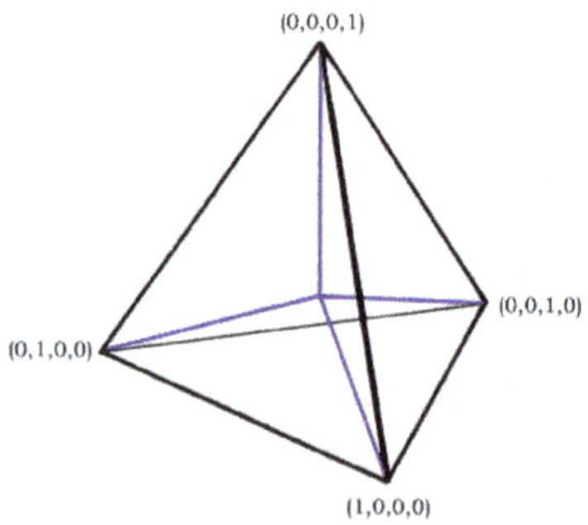

Barycentric coordinates in 3-dimensional space: four coordinates are related with one linear equation

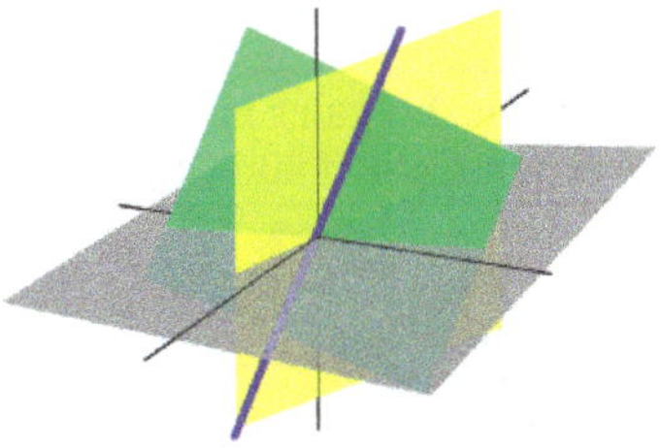

Three mutually transversal planes in the 3-dimensional space and their intersections, three lines

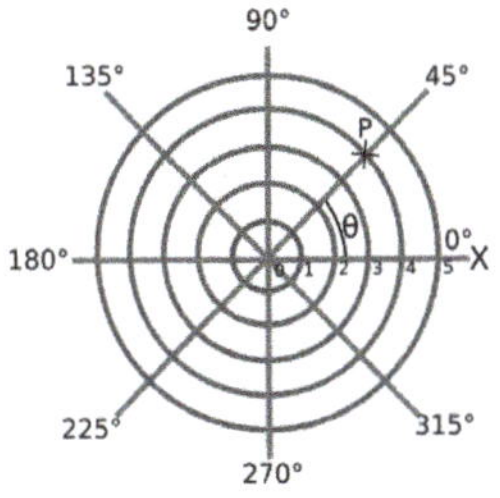

Polar coordinates

Lines, Planes, and Other Subspaces

The simplest (after points) objects in Euclidean space are flats, or Euclidean *subspaces* of lesser dimension. Points are 0-dimensional flats, 1-dimensional flats are called *(straight) lines*, and 2-dimensional flats are *planes*. $(n - 1)$-dimensional flats are called *hyperplanes*.

Any two distinct points lie on exactly one line. Any line and a point outside it lie on exactly one plane. More generally, the properties of flats and their incidence of Euclidean space are shared with affine geometry, whereas the affine geometry is devoid of distances and angles.

Line Segments and Triangles

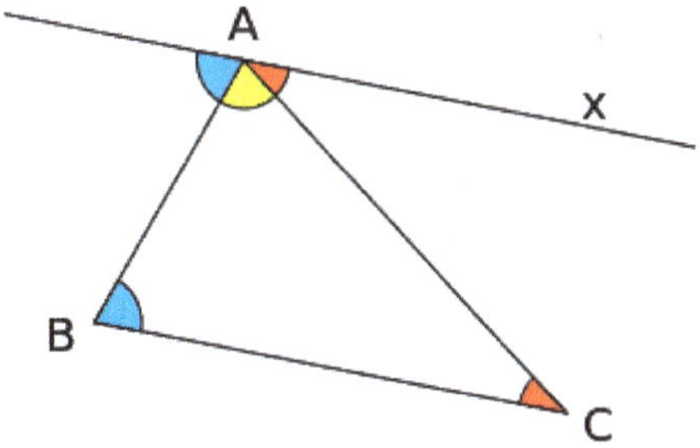

The sum of angles of a triangle is an important problem, which exerted a great influence to 19th-century mathematics. In a Euclidean space it invariably equals to 180°, or a half-turn

This is not only a line which a pair (A,B) of distinct points defines. Points on the line which lie between A and B, together with A and B themselves, constitute a line segment AB. Any line segment has the length, which equals to distance between A and B. If $A = B$, then the segment is degenerate and its length equals to 0, otherwise the length is positive.

A (non-degenerate) triangle is defined by three points not lying on the same line. Any triangle lies on one plane. The concept of triangle is not specific to Euclidean spaces, but Euclidean triangles have numerous special properties and define many derived objects.

A triangle can be thought of as a 3-gon on a plane, a special (and the first meaningful in Euclidean geometry) case of a polygon.

Polytopes and Root Systems

The Platonic solids are the five polyhedra that are most regular in a combinatoric sense, but also, their symmetry groups are embedded into O(3)	
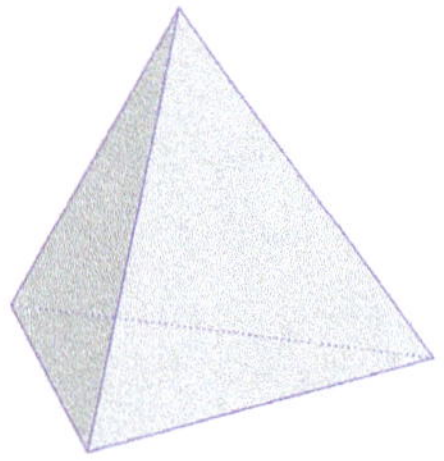 Tetrahedron	 Cube (green) and octahedron (cyan)
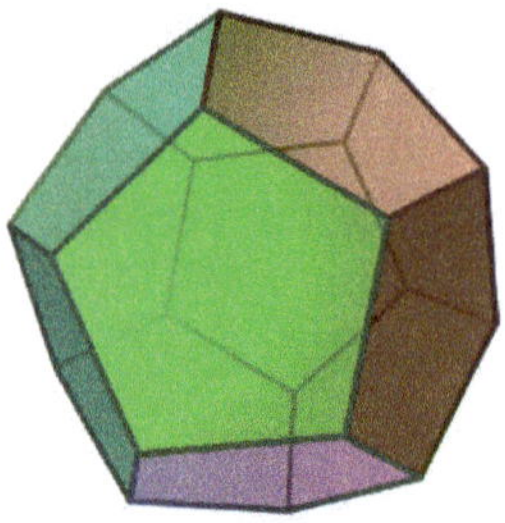 Dodecahedron	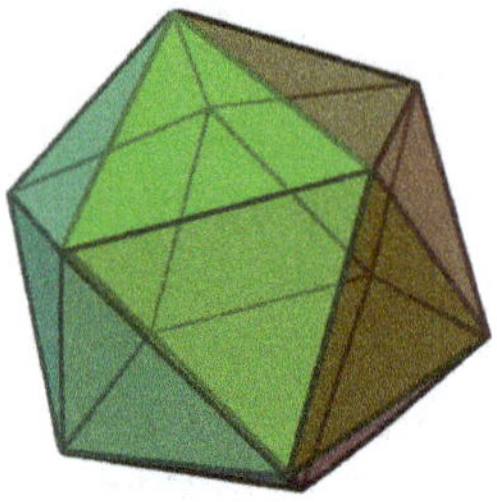 Icosahedron

Polytope is a concept that generalizes polygons on a plane and polyhedra in 3-dimensional space (which are among the earliest studied geometrical objects). A simplex is a generalization of a line segment (1-simplex) and a triangle (2-simplex). A tetrahedron is a 3-simplex.

The concept of a polytope belongs to affine geometry, which is more general than Euclidean. But Euclidean geometry distinguish *regular polytopes*. For example, affine geometry does not see the difference between an equilateral triangle and a right triangle, but in Euclidean space the former is regular and the latter is not.

Root systems are special sets of Euclidean vectors. A root system is often identical to the set of vertices of a regular polytope.

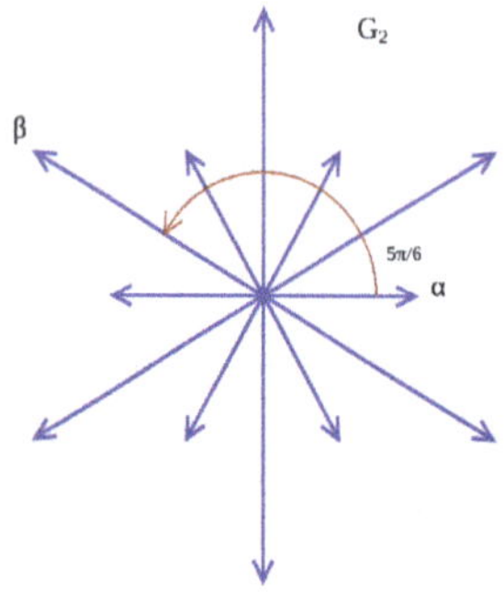

The root system G_2

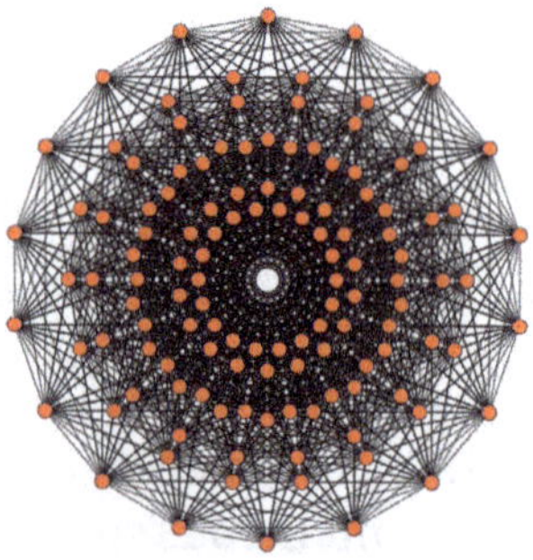

An orthogonal projection of the 2_{31} polytope, whose vertices are elements of the E_7 root system

Topology

Since Euclidean space is a metric space, it is also a topological space with the natural topology induced by the metric. The metric topology on E^n is called the Euclidean topology, and it is identical to the standard topology on R^n. A set is open if and only if it contains an open ball around each of its points; in other words, open balls form a base of the topology. The topological dimension of the Euclidean n-space equals n, which implies that spaces of different dimension are not homeomorphic. A finer result is the invariance of domain, which proves that any subset of n-space, that is (with its subspace topology) homeomorphic to an open subset of n-space, is itself open.

Applications

Aside from countless uses in fundamental mathematics, a Euclidean model of the physical space can be used to solve many practical problems with sufficient precision. Two usual approaches are a fixed, or *stationary* reference frame (i.e. the description of a motion of objects as their positions that change continuously with time), and the use of Galilean space-time symmetry (such as in Newtonian mechanics). To both of them the modern Euclidean geometry provides a convenient formalism; for example, the space of Galilean velocities is itself a Euclidean space.

Topographical maps and technical drawings are planar Euclidean. An idea behind them is the scale invariance of Euclidean geometry, that permits to represent large objects in a small sheet of paper, or a screen.

Alternatives and Generalizations

Although Euclidean spaces are no longer considered to be the only possible setting for a geometry, they act as prototypes for other geometric objects. Ideas and terminology from Euclidean geometry (both traditional and analytic) are pervasive in modern mathematics, where other geometric objects share many similarities with Euclidean spaces, share part of their structure, or embed Euclidean spaces.

Curved Spaces

A smooth manifold is a Hausdorff topological space that is locally diffeomorphic to Euclidean space. Diffeomorphism does not respect distance and angle, but if one additionally prescribes a smoothly varying inner product on the manifold's tangent spaces, then the result is what is called a Riemannian manifold. Put differently, a Riemannian manifold is a space constructed by deforming and patching together Euclidean spaces. Such a space enjoys notions of distance and angle, but they behave in a curved, non-Euclidean manner. The simplest Riemannian manifold, consisting of R^n with a constant inner product, is essentially identical to Euclidean n-space itself. Less trivial examples are n-sphere and hyperbolic spaces. Discovery of the latter in the 19th century was branded as the non-Euclidean geometry.

Also, the concept of a Riemannian manifold permits an expression of the Euclidean structure in any smooth coordinate system, via metric tensor. From this tensor one can compute the Riemann curvature tensor. Where the latter equals to zero, the metric structure is locally Euclidean (it means that at least some open set in the coordinate space is isometric to a piece of Euclidean space), no matter whether coordinates are affine or curvilinear.

Indefinite Quadratic Form

If one replaces the inner product of a Euclidean space with an indefinite quadratic form, the result is a pseudo-Euclidean space. Smooth manifolds built from such spaces are called pseudo-Riemannian manifolds. Perhaps their most famous application is the theory of relativity, where flat spacetime is a pseudo-Euclidean space called Minkowski space, where rotations correspond to motions of hyperbolic spaces mentioned above. Further generalization to curved spacetimes form pseudo-Riemannian manifolds, such as in general relativity.

Other Number Fields

Another line of generalization is to consider other number fields than one of real numbers. Over complex numbers, a Hilbert space can be seen as a generalization of Euclidean dot product structure, although the definition of the inner product becomes a sesquilinear form for compatibility with metric structure.

References

- Arkhangel'skii, A.V. (2001), "Compact space", in Hazewinkel, Michiel, Encyclopedia of Mathematics, Springer, ISBN 978-1-55608-010-4 .
- Dawson, John W. junior (1993). "The compactness of first-order logic: From Gödel to Lindström". History and Philosophy of Logic. 14: 15–37. doi:10.1080/01445349308837208.
- Kline, Morris (1972), Mathematical thought from ancient to modern times (3rd ed.), Oxford University Press (published 1990), ISBN 978-0-19-506136-9 .
- Fréchet, Maurice (1906), "Sur quelques points du calcul fonctionnel", Rendiconti del Circolo Matematico di Palermo, 22 (1): 1–72, doi:10.1007/BF03018603 .
- Ball, W.W. Rouse (1960) [1908]. A Short Account of the History of Mathematics (4th ed.). Dover Publications. pp. 50–62. ISBN 0-486-20630-0.
- Scarborough, C.T.; Stone, A.H. (1966), "Products of nearly compact spaces", Transactions of the American Mathematical Society, Transactions of the American Mathematical Society, Vol. 124, No. 1, 124 (1): 131–147, JSTOR 1994440, doi:10.2307/1994440 .
- E.D. Solomentsev (7 February 2011). "Euclidean space.". Encyclopedia of Mathematics. Springer. Retrieved 1 May 2014.
- Steen, Lynn Arthur; Seebach, J. Arthur Jr. (1995) [1978], Counterexamples in Topology (Dover Publications reprint of 1978 ed.), Berlin, New York: Springer-Verlag, ISBN 978-0-486-68735-3, MR 507446

5

An Overview of Countability Axiom

A set of axioms, along with a number of statements, called postulates which are assumed to be "true" are called axiomatic systems. An axiomatic system is said to be complete if every statement or its negative is derivable. The aspects elucidated in this chapter are of vital importance, and provide a better understanding of topology.

Axiomatic System

In mathematics, an axiomatic system is any set of axioms from which some or all axioms can be used in conjunction to logically derive theorems. A theory consists of an axiomatic system and all its derived theorems. An axiomatic system that is completely described is a special kind of formal system. A formal theory typically means an axiomatic system, for example formulated within model theory. A formal proof is a complete rendition of a mathematical proof within a formal system.

Properties

An axiomatic system is said to be *consistent* if it lacks contradiction, i.e. the ability to derive both a statement and its denial from the system's axioms.

In an axiomatic system, an axiom is called *independent* if it is not a theorem that can be derived from other axioms in the system. A system will be called independent if each of its underlying axioms is independent. Although independence is not a necessary requirement for a system, consistency is.

An axiomatic system will be called *complete* if for every statement, either itself or its negation is derivable.

Relative Consistency

Beyond consistency, relative consistency is also the mark of a worthwhile axiom system. This is when the undefined terms of a first axiom system are provided definitions from a second, such that the axioms of the first are theorems of the second.

A good example is the relative consistency of neutral geometry or absolute geometry with respect to the theory of the real number system. Lines and points are undefined terms in absolute geometry, but assigned meanings in the theory of real numbers in a way that is consistent with both axiom systems.

Models

A model for an axiomatic system is a well-defined set, which assigns meaning for the undefined terms presented in the system, in a manner that is correct with the relations defined in the system. The existence of a concrete model proves the consistency of a system. A model is called concrete if the meanings assigned are objects and relations from the real world, as opposed to an abstract model which is based on other axiomatic systems.

Models can also be used to show the independence of an axiom in the system. By constructing a valid model for a subsystem without a specific axiom, we show that the omitted axiom is independent if its correctness does not necessarily follow from the subsystem.

Two models are said to be isomorphic if a one-to-one correspondence can be found between their elements, in a manner that preserves their relationship. An axiomatic system for which every model is isomorphic to another is called categorial (sometimes categorical), and the property of categoriality (categoricity) ensures the completeness of a system.

Axiomatic Method

Stating definitions and propositions in a way such that each new term can be formally eliminated by the priorly introduced terms requires primitive notions (axioms) to avoid infinite regress. This way of doing mathematics is called the axiomatic method.

A common attitude towards the axiomatic method is logicism. In their book *Principia Mathematica*, Alfred North Whitehead and Bertrand Russell attempted to show that all mathematical theory could be reduced to some collection of axioms. More generally, the reduction of a body of propositions to a particular collection of axioms underlies the mathematician's research program. This was very prominent in the mathematics of the twentieth century, in particular in subjects based around homological algebra.

The explication of the particular axioms used in a theory can help to clarify a suitable level of abstraction that the mathematician would like to work with. For example, mathematicians opted that rings need not be commutative, which differed from Emmy Noether's original formulation. Mathematicians decided to consider topological spaces more generally without the separation axiom which Felix Hausdorff originally formulated.

The Zermelo-Fraenkel axioms, the result of the axiomatic method applied to set theory, allowed the "proper" formulation of set-theory problems and helped to avoid the paradoxes of naïve set theory. One such problem was the Continuum hypothesis. Zermelo–Fraenkel set theory with the historically controversial axiom of choice included is commonly abbreviated ZFC, where C stands for choice. Many authors use ZF to refer to the axioms of Zermelo–Fraenkel set theory with the axiom of choice excluded. Today ZFC is the standard form of axiomatic set theory and as such is the most common foundation of mathematics.

History

Mathematical methods developed to some degree of sophistication in ancient Egypt, Babylon, India, and China, apparently without employing the axiomatic method.

Euclid of Alexandria authored the earliest extant axiomatic presentation of Euclidean geometry and number theory. Many axiomatic systems were developed in the nineteenth century, including non-Euclidean geometry, the foundations of real analysis, Cantor's set theory, Frege's work on foundations, and Hilbert's 'new' use of axiomatic method as a research tool. For example, group theory was first put on an axiomatic basis towards the end of that century. Once the axioms were clarified (that inverse elements should be required, for example), the subject could proceed autonomously, without reference to the transformation group origins of those studies.

Issues

Not every consistent body of propositions can be captured by a describable collection of axioms. Call a collection of axioms recursive if a computer program can recognize whether a given proposition in the language is an axiom. Gödel's First Incompleteness Theorem then tells us that there are certain consistent bodies of propositions with no recursive axiomatization. Typically, the computer can recognize the axioms and logical rules for deriving theorems, and the computer can recognize whether a proof is valid, but to determine whether a proof exists for a statement is only soluble by "waiting" for the proof or disproof to be generated. The result is that one will not know which propositions are theorems and the axiomatic method breaks down. An example of such a body of propositions is the theory of the natural numbers. The Peano Axioms (described below) thus only partially axiomatize this theory.

In practice, not every proof is traced back to the axioms. At times, it is not clear which collection of axioms a proof appeals to. For example, a number-theoretic statement might be expressible in the language of arithmetic (i.e. the language of the Peano Axioms) and a proof might be given that appeals to topology or complex analysis. It might not be immediately clear whether another proof can be found that derives itself solely from the Peano Axioms.

Any more-or-less arbitrarily chosen system of axioms is the basis of some mathematical theory, but such an arbitrary axiomatic system will not necessarily be free of contradictions, and even if it is, it is not likely to shed light on anything. Philosophers of mathematics sometimes assert that mathematicians choose axioms "arbitrarily", but the truth is that although they may appear arbitrary when viewed only from the point of view of the canons of deductive logic, that is merely a limitation on the purposes that deductive logic serves.

Example: The Peano Axiomatization of Natural Numbers

The mathematical system of natural numbers 0, 1, 2, 3, 4, ... is based on an axiomatic system first written down by the mathematician Peano in 1889. He chose the axioms, in the language of a single unary function symbol S (short for "successor"), for the set of natural numbers to be:

- There is a natural number 0.
- Every natural number a has a successor, denoted by Sa.
- There is no natural number whose successor is 0.
- Distinct natural numbers have distinct successors: if $a \neq b$, then $Sa \neq Sb$.
- If a property is possessed by 0 and also by the successor of every natural number it is possessed by, then it is possessed by all natural numbers ("*Induction axiom*").

Axiomatization

In mathematics, axiomatization is the formulation of a system of statements (i.e. axioms) that relate a number of primitive terms in order that a consistent body of propositions may be derived deductively from these statements. Thereafter, the proof of any proposition should be, in principle, traceable back to these axioms.

Axiom of Countability

In mathematics, an axiom of countability is a property of certain mathematical objects (usually in a category) that asserts the existence of a countable set with certain properties. Without such an axiom, such a set might not provably exist.

Important Examples

Important countability axioms for topological spaces include:

- sequential space: a set is open if every sequence convergent to a point in the set is eventually in the set

- first-countable space: every point has a countable neighbourhood basis (local base)
- second-countable space: the topology has a countable base
- separable space: there exists a countable dense subset
- Lindelöf space: every open cover has a countable subcover
- σ-compact space: there exists a countable cover by compact spaces

Relationships with Each Other

These axioms are related to each other in the following ways:

- Every first countable space is sequential.
- Every second-countable space is first-countable, separable, and Lindelöf.
- Every σ-compact space is Lindelöf.
- Every metric space is first countable.
- For metric spaces second-countability, separability, and the Lindelöf property are all equivalent.

Countability Axiom

First and Second Countable Topological Spaces

Definition: A topological space $(X, \mathcal{J})$ is said to have a countable local basis (or countable basis) at a point $x \in X$ if there exists a countable collection say $\mathscr{B}_x$ of open sets containing x such that for each open set U containing x there exists $V \in \mathscr{B}_x$ with $V \subseteq U$.

Definition: A topological space $(X, \mathcal{J})$ is said to be first countable or said to satisfy the first countability axiom if for each $x \in X$ there exists a countable local base at x.

Examples: (i) Let (X, d) be a metric space then for each $x \in X$, $\mathscr{B}_x = \{B\left(x, \frac{1}{n}\right) : n \in \mathbb{N}\}$ is a countable local basis at x. Hence $(X, \mathcal{J}_d)$ is a first countable space. So, we say that every metric space (X, d) is a first countable space. (ii) Let $X = \mathbb{N}$ and $\mathcal{J} = \{\phi, X, \{1\}, \{1,2\}, \ldots, \{1,2,\ldots,n\}, \ldots,\}$ then obviously $(X, \mathcal{J})$ is a first countable topological space.

Note that this is not an interesting example of a first countable topological space. Once the topology $\mathcal{J}$ is a countable collection then $(X, \mathcal{J})$ is a first countable space.

Example: Let $X = \mathbb{R}$ and $\mathcal{J}_l$ be the lower limit topology on $\mathbb{R}$ generated by $\{[a,b) : a,b \in \mathbb{R}, a < b\}$. For each $x \in X$, $\mathscr{B}_x = \{\left[x, x+\frac{1}{n}\right) : n \in \mathbb{N}\}$ is a countable local base at x. Hence $(\mathbb{R}, \mathcal{J}_l) = \mathbb{R}_l$ is a first countable topological space. Now let us see a stronger version of first countable topological space.

Definition: If a topological space $(X, \mathcal{J})$ has a countable basis $\mathscr{B}$ then we say that $(X, \mathcal{J})$ is a second countable topological space or it satisfies the second countability axiom.

Exercise: Though it is trivial from the definition, prove that every second countable topological space $(X, \mathcal{J})$ is a first countable topological space. What about the converse?

Let X be any uncountable set and $\mathcal{J}_D$ be the discrete topology on X. Then $(X, \mathcal{J}_D)$ is first countable, but it is not second countable. In fact, for each $x \in X$, $\mathscr{B}_x = \{\{x\}\}$ is a countable local base at x. Take any open set U containing x then there exists $V = \{x\} \in \mathscr{B}_x$ such that $x \in V \subseteq U$. Hence $\mathscr{B}_x$ is a local base at x.

How to prove that $(X, \mathcal{J})$ is not a second countable topological space ? Well we use the method of proof by contradiction.

Suppose there exists a countable basis say $\mathscr{B} = \{B_1, B_2, \ldots,\}$ for $(X, \mathcal{J})$. Let us assume that each $B_k \neq \phi$. For each $k \in \mathbb{N}$, let $x_k \in B_k$. Since X is an uncountable set we can select an $x \in X$ such that $x \neq x_k$ for all $k \in \mathbb{N}$. Now $\{x\}$, the singleton set containing x, is an open set and $\mathscr{B}$ is a basis for $(X, \mathcal{J})$ implies there exists $k \in \mathbb{N}$ such that $x \in B_k \subseteq \{x\}$ this implies $B_k = \{x\}$. But $x_k \in B_k$ implies $x = x_k$, a contradiction to our assumption that $x \in X$ such that $x \neq x_k$ for all $k \in \mathbb{N}$. Hence if X is an uncountable set then the discrete topological space $(X, \mathcal{J}_D)$ is first countable but not second countable.

Also we have seen that the lower limit topological space $\mathbb{R}_l$ is first countable. Now let us prove that $\mathbb{R}_l = (\mathbb{R}, \mathcal{J}_l)$ is not a second countable topological space. That is we will have to prove that if $\mathscr{B}$ is a basis for $(\mathbb{R}, \mathcal{J}_l)$ then $\mathscr{B}$ is not a countable collection. So, fix a basis say $\mathscr{B}$ for $(\mathbb{R}, \mathcal{J}_l)$. For each $x \in \mathbb{R}$, $[x, x+1) \in \mathcal{J}_l$. Hence $\mathscr{B}$ is a basis for $(\mathbb{R}, \mathcal{J}_l)$ implies there exists $B_x \in \mathscr{B}$ such that $x \in B_x \subseteq [x, x+1)$.

For $x, y \in \mathbb{R}$, $x \neq y$ we have $[x, x+1) \neq [y, y+1)$. Also $B_x \subseteq [x, x+1)$ implies $\inf B_x \geq \inf[x, x+1) = x$. Also $x \in B_x$ implies $x \geq \inf B_x$. Hence $x = \inf B_x$. Now define $f : \mathbb{R} \to \mathscr{B}$

as $f(x) = B_x$. Then $x \neq y$ implies $B_x \neq B_y$. ($B_x = B_y$ implies inf B_x = inf B_y) That is $f(x) \neq f(y)$. Hence f is an one-one function. This implies that $f:\mathbb{R} \to f(\mathbb{R}) \subseteq \mathcal{B}$ is a bijective function. Therefore $f(\mathbb{R})$ is an uncountable set and hence $\mathcal{B}$ is an uncountable set. We have proved that if $\mathcal{B}$ is a basis for $(\mathbb{R}, \mathcal{J}_l)$ then $\mathcal{B}$ is an uncountable set. Hence $(\mathbb{R}, \mathcal{J}_l)$ cannot have a countable basis and therefore $(\mathbb{R}, \mathcal{J}_l)$ is not a second countable topological space.

It is a simple exercise to check $\overline{\mathbb{Q}} = \mathbb{R}$ in $(\mathbb{R}, \mathcal{J}_l)$. That is $\mathbb{Q}$ is a countable dense subset of $\mathbb{R}$ with respect to $(\mathbb{R}, \mathcal{J}_l)$. Such a topological space is known as a separable topological space.

Definition: A topological space $(X, \mathcal{J})$ is said to be a separable topological space if there exists a countable subset say A of X such that $\overline{A} = X$.

Definition: A topological space $(X, \mathcal{J})$ is said to be a Lindelöf space if for any collection $\mathcal{A}$ of open sets such that $X = \bigcup_{A \in \mathcal{A}} A$, there exists a countable subcollection say $\mathcal{B} \subseteq \mathcal{A}$ such that $X = \bigcup_{B \in \mathcal{B}} B$. That is, a topological space $(X, \mathcal{J})$ is said to be a Lindelöf space if and only if every open cover of X has a countable subcover for X.

By definition every compact topological space $(X, \mathcal{J})$ is a Lindelöf space. But the converse need not be true. It is easy to prove that $\mathbb{R}$ ($\mathbb{R}$ with usual topology) is a Lindelöf space. But $\mathbb{R}$ is not compact space.

Now let us prove that every second countable topological space is a Lindelöf space.

Theorem 1. If $(X, \mathcal{J})$ is a second countable topological space then $(X, \mathcal{J})$ is a Lindelöf space.

Proof: Let $\mathcal{B} = \{B_1, B_2, B_3, \ldots\}$ be a countable basis for $(X, \mathcal{J})$ and $\mathcal{A}$ be an open cover for X. Let us assume that, $X \neq \phi$, $A \neq \phi$ for each A ∈ $\mathcal{A}$ and $B \neq \phi$, for each $B \in \mathcal{B}$. Fix A ∈ $\mathcal{A}$ and x ∈ A. Now x ∈ A, A is an open set implies there exists $B \in \mathcal{B}$ such that

$$x \in B \subseteq A. \tag{1}$$

For each $n \in \mathbb{N}$, let $\mathcal{F}_n = \{A \in \mathcal{A} : B_n \subseteq A\}$. Here it is possible that $\mathcal{F}_n = \phi$, for Some $n \in \mathbb{N}$. At the same time note that, from Eq. (1), $\{n \in \mathbb{N} : \mathcal{F}_n = \phi\}$ is a nonempty set. Let $\{n \in \mathbb{N} : \mathcal{F}_n = \phi\} = \{n_1, n_2, \ldots, n_k \ldots,\}$ (it may be a finite set) and for each such k take $A_{n_k} \in \mathcal{F}_{n_k}$. This will give us $B_{n_k} \subset A_{n_k} \in \mathcal{A}$.

Let us prove that $\bigcup_{k=1}^{\infty} A_{n_k} = X$. So, let $x \in X$. Now $\mathcal{A}$ is an open cover for X implies $x \in A$ for some A. Now $x \in A$, $\mathcal{B}$ is a basis for $(X, \mathcal{J})$ implies there exists $k \in \mathbb{N}$ such that $x \in B_{n_k} \subseteq A$. This implies that $A \in \mathcal{F}_{n_k}$. Also $A_{n_k} \in \mathcal{F}_{n_k}$. Hence by our definition of $\mathcal{F}_{n_k}$, $B_{n_k} \subseteq A_{n_k}$. Hence $x \in X$ implies $x \in A_{n_k}$, for some $k \in \mathbb{N}$.

This implies that $X \subseteq \bigcup_{k=1}^{\infty} A_{n_k}$. That is $\{A_{n_k}\}_{k=1}^{\infty}$ is a countable subcover for $\mathcal{A}$. Therefore every open cover $\mathcal{A}$ of X has a countable subcover. Hence $(X, \mathcal{J})$ is a Lindelöf space.

Note. Recall that, for $1 \le p < \infty, l_p = \left\{ x = (x_n)_{n=1}^{\infty} : \sum_{n=1}^{\infty} |x_n|^p < \infty \right\}$ is a second countable metric space, where $x = (x_n) \in l_p$, $y = (y_n) \in l_p$,

$$d_p((x_n),(y_n)) = d_p(x, y) = \left(\sum_{n=1}^{\infty} |x_n - y_n|^p \right)^{\frac{1}{p}}.$$

Also note that $\mathbb{R}^n = \{x = (x_1, x_2, \ldots, x_n) : x_i \in \mathbb{R}, i = 1, 2, \ldots, n\}$ is a second countable metric space with respect to any of the metric given by $d_p(x, y) = \left(\sum_{n=1}^{\infty} |x_n - y_n|^p \right)^{\frac{1}{p}}$ for $x = (x_1, x_2, \ldots, x_n) \in \mathbb{R}^n$, $y = (y_1, y_2, \ldots, y_n) \in \mathbb{R}^n, 1 \le p < \infty$ or $d_\infty(x, y) = max\{|x_k - y_k| : k = 1, 2, \ldots, n\}$. So, all the above mentioned metric spaces are all Lindelöf spaces. But none of these metric spaces is a compact space. Now let us prove that a second countable topological space is a separable space.

Theorem 2. Every second countable topological space $(X, \mathcal{J})$ is a separable space.

Proof: Given that $(X, \mathcal{J})$ is a second countable topological space. Hence there exists a countable basis say $\mathcal{B} = \{B_1, B_2, \ldots\}$ for $(X, \mathcal{J})$. When we write $\mathcal{B} = \{B_1, B_2, \ldots\}$, it does not mean that $\mathcal{B}$ is a countably infinite set. It means that either for some $n \in \mathbb{N}$, $\mathcal{B} = \{B_1, B_2, \ldots, B_n\}$ or $\mathcal{B} = \phi$ or $\mathcal{B}$ is a countably infinite set. If $X \neq \phi$ then $\mathcal{B} \neq \phi$. If for some $k \in \mathbb{N}$, $B_k = \phi$, then $\mathcal{B}' = \{B_1, B_2, \ldots, B_{k-1}, B_{k+1}, \ldots, \}$ is also a basis for $(X, \mathcal{J})$.

So, let us assume that each $B_n \neq \phi$ for all n. Since $B_n \neq \phi$, for each $n \in \mathbb{N}$, let $x_n \in B_n$ (note that by axiom of choice there exists a function $f : \mathbb{N} \to \bigcup_{n=1}^{\infty} B_n$ such that $x_n = f(n) \in B_n$) and $A = \{x_1, x_2, x_3, \ldots, \}$. Here also it is quite possible that A is a finite set. Now let us prove that $\overline{A} = X$. So, take an $x \in X$ and an open set U containing x. Now $\mathcal{B}$ is a basis for $(X, \mathcal{J})$, U is an open set containing x implies there exists $B_n \in \mathscr{B}$ such that $x \in B_n$ and $B_n \subseteq U$. Also $x_n \in B_n$. Hence $x_n \in U \cap A$. This gives that $U \cap A \neq \phi$. That is we have proved that $U \cap A \neq \phi$ for each open set U containing x. Hence $x \in \bar{A}$. That is $x \in X$ and hence $x \in \bar{A}$ and hence $\bar{A} = X$. Therefore $(X, \mathcal{J})$ has a countable dense subset and therefore $(X, \mathcal{J})$ is a separable space.

Now let us prove that subspace of a separable metric space is separable.

Theorem 3. Let (X, d) be a separable metric space and Y be a subspace of X (that is $Y \subseteq X$, and for $x, y \in Y$, $d_Y(x, y) = d(x, y)$). Then (Y, d_Y) is a separable space.

Proof: (X, d) is a separable metric space implies there exists a countable subset say $A = \{x_1, x_2, x_3, \ldots,\}$ of X such that $\bar{A} = X$ (here $\bar{A}$ denotes the closure of A with respect to (X, d)). We will have to find a countable subset say B of Y such that $\bar{B}_Y = Y$ (here $\bar{B}_Y = \bar{B} \cap Y$, the closure of B with respect to the subspace (Y, d_Y)). For $n \in \mathbb{N}$, let

$$A_{n,k} = B\left(x_n, \frac{1}{k}\right) \cap Y.$$

Here we do not know whether $A_{n,k} = \phi$ or $A_{n,k} \neq \phi$. If $A_{n,k} \neq \phi$ $(n, k \in \mathbb{N})$ let $a_{n,k} \in A_{n,k}$ be a fixed element. Then $B = \{a_{n,k} : a_{n,k} \in A_{n,k} \text{ whenever } A_{n,k} \neq \phi\}$ is a countable subset of Y. Now let us prove that $\bar{B}_Y = \bar{B} \cap Y = Y$. So let $x \in Y$ and U be an open set in X containing x. Hence there exists $k \in \mathbb{N}$ such that $B\left(x, \frac{1}{k}\right) \subseteq U$. Again $x \in \bar{A} = X$ implies $B\left(x, \frac{1}{2k}\right) \cap A \neq \phi$. Then there exists $x_n \in A$ such that $x_n \in B\left(x, \frac{1}{2k}\right)$.

Therefore $x \in B\left(x_n, \frac{1}{2k}\right) \cap Y = A_{n,2k}$. Hence $A_{n,2k} \neq \phi$. Now $A_{n,2k} \neq \phi$ implies $a_{n,2k} \in B$.

Further $a_{n,2k} \in B\left(x, \frac{1}{2k}\right)$. Now $d\left(x, a_{n,2k}\right) \leq d\left(x, x_n\right) + d\left(x_n, a_{n,2k}\right) < \frac{1}{2k} + \frac{1}{2k} = \frac{1}{k}$.

Hence $a_{n,2k} \in B\left(x -\right) \cap B \subseteq U \cap B$. That is $U \cap B \neq \phi$ for each open set U containing x. This implies $x \in \bar{B}$. Also $x \in Y$. Therefore $x \in \bar{B} \cap Y = \bar{B}_Y$. This implies $Y \subseteq \bar{B}_Y \subseteq Y$ and hence $\bar{B}_Y = Y$. That is B is a countable dense subset of Y. This proves that the subspace (Y, d_Y) is a separable metric space.

Subspace of a Separable Topological Space need not be Separable

We give an example to show that subspace of a separable topological space need not be separable.

Let $X = \{(x, y) : x \in \mathbb{R}, y \geq 0\}$. Basic open sets are of the type: (i) for $(x, y) \in \mathbb{R}^2$, $x \in \mathbb{R}$, $y > 0$ basic open sets containing (x, y) are of the form $B((x, y), r)$, $o < r < y$, and (ii) for $(x, 0) \in \mathbb{R}^2$, $x \in \mathbb{R}$, basic open sets are of the form $(B(x, o), r) \cap X) \setminus \{(y, o) : o < |y - x| < r\}$, $r > o$. Here $B((x, y), r) = \{(a, b) \in \mathbb{R}^2 : d((x, y), (a, b)) = \sqrt{(x-a)^2 + (y-b)^2} < r\}$, the open ball centered at (x, y) and radius r with respect to the Euclidean metric d on $\mathbb{R}^2$.

It is easy to see that the above collection of sets will form a basis for a topology on X.

Let $\mathcal{J}$ be the topology on X induced by the collection of basic open sets described as above and if $A = \{(x, y) : x, y \in \mathbb{Q}, y \geq 0\}$ then A is a countable subcollection of X such that $\overline{A} = X$. That is A is a countable dense subset of X, and hence $(X, \mathcal{J})$ is a separable topological space.

Now for $Y = \{(x, 0) : x \in \mathbb{R}\}$ (an uncountable set), $\mathcal{J}_Y = \mathcal{P}(Y)$. That is the subspace $(Y, \mathcal{J}_Y)$ of $(X, \mathcal{J})$ is the discrete topological space. That is every subset U of Y is both open and closed in $(Y, \mathcal{J}_Y)$. Therefore if B is a countable subset of Y then $\overline{B} = B \neq Y$ here ($\overline{B} = \overline{B}_{\mathcal{J}_Y}$, the is closure of B in $(Y, \mathcal{J}_Y)$). This proves that $(Y, \mathcal{J}_Y)$ is not a separable subspace, though $(X, \mathcal{J})$ is a separable topological space.

Note that we have already proved that every second countable topological space is separable.

Exercise:. Prove that every separable metric space is second countable.

Exercise: Prove that $\mathbb{R}^n = \{(x_1, x_2, \ldots, x_n) : x_i \in \mathbb{R}, i = 1, 2, \ldots, n\}$ is a separable metric space for $1 \leq p < \infty$, $d_p(x, y) = \left(\sum_{i=1}^{n} |x_i - y_i|^p\right)^{\frac{1}{p}}$, $x = (x_1, x_2, \ldots, x_n) \in \mathbb{R}^n$, $y = (y_1, y_2, \ldots, y_n) \in \mathbb{R}^n$ and $d_\infty(x, y) = max\{|x_i - y_i| : 1 \leq i \leq n\}$.

It is easy to prove that if $\mathcal{J}_2$ is the topology induced by d_2 then $\mathcal{J}_p = \mathcal{J}_2, \forall p \geq 1$. That is all these metrics d_p, $1 \leq p \leq \infty$ will induce the same topology on $\mathbb{R}^n$. So if we want to prove that $(\mathbb{R}^n, d_p)$ is a separable metric space, it is enough to prove that $(\mathbb{R}^n, \mathcal{J}_1)$ (or say $(\mathbb{R}^n, \mathcal{J}_2)$) is separable.

For $1 \leq p < \infty$, let $l_p = \{x = (x_n) : x_n \in \mathbb{R}$ for all n and $\sum_{n=1}^{\infty} |x_n|^p < \infty\}$. If we define, $d_p(x, y) = \left(\sum_{n=1}^{\infty} |x_n - y_n|^p\right)^{\frac{1}{p}}$, then d_p is a metric on l_p. Now let us see how to prove that

(l_p, d_p) is a separable metric space.

Step 1: For each $n \in \mathbb{N}$, let $A_n = \{(r_1, r_2, \ldots, r_n, \ldots, 0, 0, \ldots) : r_i \in \mathbb{Q},\ i = 1, 2, \ldots, n\}$.

If we define $f : \mathbb{Q} \times \mathbb{Q} \times \cdots \times \mathbb{Q}\ (n\ times) = \mathbb{Q}^n \to A_n\ as\ f(r_1, r_2, \ldots, r_n) = (r_1, r_2, \ldots, r_n, \ldots, 0, 0, \ldots)$. Then f is a bijective function. Now $\mathbb{Q}$ is a countable set implies $\mathbb{Q} \times \mathbb{Q} \times \cdots \times \mathbb{Q}$ (finite product of countable sets is countable) is countable. Hence there is a bijection between $\mathbb{Q}^n$ and A_n implies A_n is a countable set. Now each A_n is a countable set implies $\bigcup_{n=1}^{\infty} A_n$ is also countable.

We leave it as an exercise to prove that $\overline{\bigcup_{n=1}^{\infty} A_n} = l_p$. That is $\bigcup_{n=1}^{\infty} A_n$ is a countable dense subset of l_p. Hence (l_p, d_p) is a separable metric space. These spaces are important examples of Banach spaces. If $l_\infty = \{x = (x_n) : (x_n)$ is a bounded real sequence$\}$ and $d_\infty(x, y) = \sup\{|x_n - y_n| : n \geq 1\}$, then (l_∞, d_∞) is also a metric space. Let $X = \{x = (x_n) : x_n = 0$ or $x_n = 1\}$. For $x, y \in X,\ x \neq y,\ d(x, y) = 1$. Hence (X, d_∞) (that is d_∞ is restricted to the subspace X of l_∞) is a metric space. Now the topology $\mathcal{J}$ on X induced by the metric d_∞ is the discrete topology on X. In this topological space $(X, \mathcal{J})$ every subset A of X is both open and closed. Therefore, if A is a countable subset of X, then $\overline{A} = A \neq X$, (note that X is an uncountable subset of X). Now the subspace X of l_∞ is not a separable space implies l_∞ is not a separable space.

Properties of First Countable Topological Spaces

Theorem 4. If $(X, \mathcal{J})$ is a first countable topological space then for each $x \in X$ there exists a countable local base say $\{V_n(x)\}_{n=1}^{\infty}$ *such that* $V_{n+1}(x) \subseteq V_n(x)$.

Proof: Fix $x \in X$. Now $(X, \mathcal{J})$ is a first countable topological space implies there exists a countable local base say $\{U_n\}_{n=1}^{\infty}$ at x. Let $V_n(x) = U_1 \cap U_2 \cap \cdots \cap U_n$ then $\{V_n(x)\}_{n=1}^{\infty}$ is a collection of open sets such that $V_{n+1}(x) \subseteq V_n(x)$ for all $n \in \mathbb{N}$. So, it is enough to prove that $\{V_n(x)\}_{n=1}^{\infty}$ is a local base at x. So start with an open set V containing x. Now $\{U_n\}_{n=1}^{\infty}$ is a local base at x and V is an open set containing x implies there exists $n_0 \in \mathbb{N}$ such that $U_{n_0} \subseteq V$. By the definition of $V_n(x)$'s we have $V_{n_0}(x) \subseteq U_{n_0}$. Hence we have the following: for each open set V containing x there exists $n_0 \in \mathbb{N}$ such that $V_{n_0}(x) \subseteq V$. This implies that $\{V_n(x)\}$ is a local base at x satisfying $V_{n+1}(x) \subseteq V_n(x)$ for all $n \in \mathbb{N}$.

Let us use the above characterization of a first countable base to show that, in some sense, first countable topological spaces behave like metric spaces.

Theorem 5. Let $(X, \mathcal{J})$ be a first countable topological space and A be a nonempty subset of X. Then for each $x \in X$, $x \in \overline{A}$ if and only if there exists a sequence $\{x_n\}_{n=1}^{\infty}$ in A such that $x_n \to x$ as $n \to \infty$.

Proof: First let us assume that $x \in \overline{A}$. Now $(X, \mathcal{J})$ is a first countable topological space implies there exists a countable local base say $\mathcal{B} = \{V_n\}_{n=1}^{\infty}$ such that $V_{n+1} \subseteq V_n$, for all $n \in \mathbb{N}$. Hence $x \in \overline{A}$ implies $A \cap V_n \neq \phi$, for each $n \in \mathbb{N}$. Let $x_n \in A \cap V_n$.

Claim: $x_n \to x$ as $n \to \infty$.

So start with an open set U containing x (enough to start with V_n) then there exists $n_0 \in \mathbb{N}$ such that $x \in V_{n_0} \subseteq U$. Hence $x_n \in V_n \subseteq V_{n_0} \subseteq U$ for all $n \geq n_0$. That is $x_n \in U$ for all $n \geq n_0$. This means $x_n \to x$ as $n \to \infty$.

Conversely, suppose there exists a sequence $\{x_n\}_{n=1}^{\infty}$ in A such that $x_n \to x$. Then for each open set U containing x there exists a positive integer n_0 such that $x_n \in U$ for all $n \geq n_0$. In particular $x_{n_0} \in U \cap A$. Hence for each open set U containing x, $U \cap A \neq \phi$ and this implies $x \in \overline{A}$.

Theorem 6. Let X and Y be topological spaces and further suppose X is a first countable topological space. Then a function $f : X \to Y$ is continuous at a point $x \in X$ if and only if for every sequence $\{x_n\}_{n=1}^{\infty}$ in X, $x_n \to x$ as $n \to \infty$, then the sequence $\{f(x_n)\}_{n=1}^{\infty}$ converges to $f(x)$ in Y.

Proof: Assume that $f : X \to Y$ is continuous at a point $x \in X$. Also assume that $\{x_n\}_{n=1}^{\infty}$ is a sequence in X such that $x_n \to x$ as $n \to \infty$.

To prove: $f(x_n) \to f(x)$ in Y.

So start with an open set V in Y containing $f(x)$. Since f is continuous at x, $U = f^{-1}(V)$ is an open set in X. Now $f(x) \in V$ implies $x \in f^{-1}(V) = U$. That is U is an open set containing x. Hence $x_n \to x$ implies there exists $n_0 \in \mathbb{N}$ such that $x_n \in U$ for all $n \geq n_0$. This implies $f(x_n) \in V$ for all $n \geq n_0$. That is, whenever $x_n \to x$ as $n \to \infty$ then $f(x_n) \to f(x)$ as $n \to \infty$.

Conversely, suppose that $\{x_n\}$ is a sequence in X, $x_n \to x$ as $n \to \infty$ implies $f(x_n) \to f(x)$. Now we will have to prove that f is continuous at x. It is to be noted that to prove this converse part we will make use of the fact that X is a first countable space. Now X is a first countable space implies there exists a local base $\{V_n\}_{n=1}^{\infty}$ at x such that $V_{n+1} \subseteq V_n$ for all $n \in \mathbb{N}$. We will use the method of proof by contradiction. If f is not continuous at x then there should exist an open set W containing $f(x)$ such that $f(U) \nsubseteq W$ for any open set U containing x. In particular for such an open set W, $f(V_n) \nsubseteq W$ for all n = 1, 2, 3, ... Hence there exists $x_n \in V_n$ such that $f(x_n) \notin W$.

Claim: $x_n \to x$ as $n \to \infty$. So start with an open set V in X containing x.

Now $\{V_n\}_{n=1}^{\infty}$ is a local base at x implies there exists $n_0 \in \mathbb{N}$ such that $V_{n_0} \subseteq V$. Hence $x_n \in V_n \subseteq V_{n_0} \subseteq V$ for all n ≥ n_0. That is for each open set V containing x there exists $n_0 \in \mathbb{N}$ such that $x_n \in V$ for all n ≥ n_0. Hence $x_n \to x$ as $n \to \infty$. But this sequence $\{x_n\}$ in X is such that $f(x_n) \notin W$, where W is an open set containing $f(x)$. So we have arrived at a contradiction to our assumption namely $x_n \in X$, $x_n \to x \in X$ implies $f(x_n) \to f(x)$. We arrived at this contradiction by assuming f is not continuous at x. Therefore our assumption is wrong and hence f is continuous at x.

Example: Let $J_c = \{A \subseteq \mathbb{R} : A^c \text{ is countable or } A^c = \mathbb{R}\}$, the co-countable topology on $\mathbb{R}$, and $X = (\mathbb{R}, J_c)$, $Y = (\mathbb{R}, J_s)$, where J_s is the standard topology on $\mathbb{R}$. Define $f : X \to Y$ as $f(x) = x$ for all $x \in X$. Suppose $\{x_n\}$ is a sequence in X such that $\{x_n\}$ converges to $x \in X = \mathbb{R}$. Then it is easy to prove that there exists $n_0 \in \mathbb{N}$ such that $x_n = x$ for all n ≥ n_0. (If this statement is not true then there exists a subsequence $\{x_{n_k}\}_{k=1}^{\infty}$ of $\{x_n\}_{n=1}^{\infty}$ such that $x_{n_k} \neq x$ for all $k \in \mathbb{N}$. Then $U = \mathbb{R} \setminus \{x_{n_k} : k \in \mathbb{N}\}$ is an open set in X containing x. Hence $\{x_n\}$ converges to x in X implies there exists $n_0 \in \mathbb{N}$ such that $x_n \in U$ for all n ≥ n_0. In particular for $k \geq n_o$, $n_k \geq k \geq n_o$ and this implies $x_{n_k} \in U$.) So we have the following: $x_n \to x$ in X implies $f(x_n) \to f(x)$ in Y. But the given function $f : X \to Y$ is not a continuous function (note: $f^{-1}(0, 1) = (0, 1)$ is not an open set in $(\mathbb{R}, J_c)$). This example does not give any contradiction to theorem 6. From this example we conclude that $X = (\mathbb{R}, J_c)$ is not a first countable topological space.

Regular and Normal Topological Spaces

Definition: A topological space (X, J) is called a T_1 space if for each $x \in X$, the singleton set $\{x\}$ is a closed set in (X, J).

Definition: A T_1-topological space (X, J) is called a regular space if for each $x \in X$ and for each closed subset A of X with $x \notin A$, there exist open sets U, V in X satisfying the following:

(i) $x \in U, A \subseteq V$, (ii) $U \cap V = \phi$.

Result: Every regular topological space (X, J) is a Hausdorff space.

Proof: Let $x, y \in X$, $x \neq y$. By definition every regular space is a T_1-space. Hence $\{y\}$ is a closed set. Also $x \neq y$ implies $x \notin A = \{y\}$. Now $\{y\}$ is a closed set which does not contain x. Since (X, J) is a regular space, there exist open sets U, V in X satisfying the following:

(i) $x \in U, A = \{y\} \subseteq V$,

(ii) $U \cap V = \phi$ that is U, V are open sets in X such that $x \in U$, $y \in V$ and $U \cap V = \phi$. Hence $(X, \mathcal{J})$ is a Hausdorff topological space.

Exercise: Prove that every Hausdorff space is a T_1-space.

Example: Let X be an infinite set and $\mathcal{J}_f$ be the cofinite topology on X. Then $(X, \mathcal{J}_f)$ is a T_1-space, but $(X, \mathcal{J}_f)$ is not a Hausdorff space. For each $x \in X$, $U = X\backslash\{x\}$ is an open set. Hence $U^c = X\backslash U = \{x\}$ is a closed set in X. That is for each $x \in X$, the singleton set $\{x\}$ is a closed set. Therefore $(X, \mathcal{J})$ is a T_1 - space. Take any $x, y \in X, x \neq y$. Suppose there exist open sets U, V in X such that $x \in U$, $y \in V$ and $U \cap V = \phi$. Now U, V are nonempty open subsets of the cofinite topological space $(X, \mathcal{J}_f)$ implies U^c, V^c are finite sets. Hence $X = \phi^c = (U \cap V)^c = U^c \cup V^c$ is a finite set. Therefore there cannot exist any open sets U, V in $(X, \mathcal{J}_f)$ satisfying $x \in U$, $y \in V$ and $U \cap V = \phi$. This means $(X, \mathcal{J}_f)$ is not a Hausdorff space.

Now let us give an example of a topological space which is Hausdorff but not regular. Take $X = \mathbb{R}$ and $\mathscr{B}_K = \{(a, b), (a, b)\backslash K : a, b \in \mathbb{R}, a < b\}$, where $K = \left\{1, \frac{1}{2}, \frac{1}{3}, \ldots\right\}$. Now it is easy to prove that (left as an exercise) $\mathscr{B}_K$ is a basis for a topology on $\mathbb{R}$. Let $\mathcal{J}_K$ be the topology on $\mathbb{R}$ generated by $\mathscr{B}_K$. If $\mathcal{J}$ is the usual topology on $\mathbb{R}$ then we know that $\mathcal{J}$ is generated by $\mathscr{B} = \{(a, b) : a, b \in \mathbb{R}, a < b\}$. Since we have $\mathscr{B} \subseteq \mathscr{B}_K$ and this implies that $\mathcal{J} = \mathcal{J}_{\mathscr{B}} \subseteq \mathcal{J}_{\mathscr{B}_K} = \mathcal{J}_K$.

From this, it is clear that $(\mathbb{R}, \mathcal{J}_K)$ is a Hausdorff space. For $x, y \in \mathbb{R}, x \neq y$, $(\mathbb{R}, \mathcal{J})$ is a Hausdorff space implies there exist open sets U and V in $(\mathbb{R}, \mathcal{J})$ such that $x \in U$, $y \in V$ and $U \cap V = \phi$. But $\mathcal{J} \subseteq \mathcal{J}_K$. Hence $U, V \in \mathcal{J}_K$ are such that $x \in U$, $y \in V$ and $U \cap V = \phi$ and this shows that $(X, \mathcal{J}_K)$ is a Hausdorff topological space.

Is $K = \left\{1, \frac{1}{2}, \frac{1}{3}, \ldots\right\}$ a closed set? Here K is a subset of $\mathbb{R}$ and $\mathcal{J}, \mathcal{J}_K$ are two different topologies on $\mathbb{R}, 0 \in \overline{K}$ and $0 \notin K$ with respect to $(\mathbb{R}, \mathcal{J})$. Hence K is not a closed set in $(\mathbb{R}, \mathcal{J})$. But $\mathbb{R} \backslash K = \bigcup_{n=1}^{\infty} A_n$, where $A_n = (-n, n)\backslash K$ for each $n \in \mathbb{N}$. Each A_n is an open set in $(\mathbb{R}, \mathcal{J}_K)$ implies $\mathbb{R} \backslash K$ is an open set in $(\mathbb{R}, \mathcal{J}_K)$. This implies K is a closed set in $(\mathbb{R}, \mathcal{J}_K)$. Also $0 \notin K$. What are the open sets containing K? If V is an open set containing K, then for each $n \in \mathbb{N}, \frac{1}{n} \in V$, there exists a basic open set say (a_n, b_n) such that $\frac{1}{n} \in (a_n, b_n) \subseteq V$ ($\frac{1}{n} \notin (a_n, b_n)\backslash K$) and $o < a_n < b_n$ implies $K \subseteq \bigcup_{k=1}^{\infty}(a_k, b_k) \subseteq V$.

Suppose U, V are open sets such that o ∈ U and $K \subseteq V$. Since o ∈ U, there exists a basic open set B such that o ∈ B ⊆ U. If B is of the form (a, b) then $(a,\ b)\cap K \neq \phi$. So $U \cap V = \phi$. If B is of the form (a, b)\K, choose $n_0 \in \mathbb{N}$ such that $\frac{1}{n_0} < b$. Since $\frac{1}{n_0} \in V$, there exists an open interval (c, d) such that $\frac{1}{n_0} \in (c,d) \subseteq V$. Now since (a, b) ∩ (c, d) is not empty (it contains $\frac{1}{n_0}$), it is an interval and hence uncountable. As K is countable, $((a,\ b)\cap(c,\ d))\setminus K\ \neq \phi$, i.e, $((a,\ b)\setminus K)\cap(c,\ d)\ \neq \phi$. Therefore $U \cap V = \phi$.

So we have proved that there cannot exist open sets U, V in $(\mathbb{R},\ J_K)$ with o ∈ U, K ⊆ V and $U \cap V = \phi$. This shows that $(\mathbb{R},\ J_K)$ is not a regular space.

Definition: A topological space $(X,\ J)$ is said to be a normal space if and only if it satisfies:

(i) $(X,\ J)$ is a T_1-space,

(ii) A, B closed sets in X, $A \cap B = \phi$ implies there exist open sets U, V in X such that A ⊆ U, B ⊆ V and $U \cap V = \phi$.

Remark: It is to be noted that every normal space is a regular space.

Theorem 7. Every metric space (X, d) is a normal space, That is if J_d is the topology induced by the metric then the topological space $(X,\ J_d)$ is a normal space.

Proof: Let A, B be disjoint closed subsets of X. Then for each $a \in A,\ a \notin B = \overline{B}$ implies d(a, B) = inf{d(a, b) : b ∈ B} > 0. If r_a = d(a, B) > 0 then $B(a,\ r_a)\cap B = \phi$ (if there exists b_0 ∈ B such that d(b_0, a) < r_a, then r_a = d(a, B) ≤ d(a, b_0) < r_a a contradiction). Similarly for each b ∈ B there exists r_b > 0 such that $B(b,\ r_b)\cap A = \phi$. Let $U = \bigcup_{a\in A} B\left(a, \frac{r_a}{3}\right), V = \bigcup_{b \in B} B\left(a, \frac{r_b}{3}\right)$. Now it is easy to prove that $U \cap V = \phi$. Hence if A, B are disjoint closed subsets of X then there exist open sets U, V in X such that A ⊆ U, B ⊆ V and $U \cap V = \phi$. This implies $(X,\ J_d)$ is a normal space.

Theorem 8. A T_1-topological space $(X,\ J)$ is regular if and only if whenever x is a point of X and U is an open set containing x then there exists an open set V containing x such that $\overline{V} \subseteq U$.

Proof: Assume that $(X,\ J)$ is a regular topological space, $x \in X$ and U is an open set containing x. Now $x \in U$ implies $x \notin A = U^c = X \setminus U$, the complement of the open set U. Now A is a closed set and $x \notin A$. Hence X is a regular space implies there exist open sets V and W of X such that $x \in V$, $A = U^c \subseteq W$ and $V \cap W = \phi$. Now $V \cap W = \phi$ implies $V \subseteq W^c \subseteq U$ (we have $U^c \subseteq W$), $V \subseteq W^c$ implies $\overline{V} \subseteq \overline{W^c} = W^c$ (W is an open set implies

W^c is a closed set) implies $\overline{V} \subseteq U$. Hence for $x \in X$ and for each open set U containing x, there exists an open set V containing x such that $\overline{V} \subseteq U$.

Now let us assume that the above statement is satisfied. Our aim here is to prove that $(X, \mathcal{J})$ is a regular space. So take a closed set A of X and a point $x \in X \setminus A$. Now A is a closed subset of X implies $U = X \setminus A$ is an open set containing x. Hence by our assumption there exists an open set V containing x such that $\overline{V} \subseteq U = A^c$. Now $\overline{V} \subseteq A^c$ implies $A \subseteq \left(\overline{V}\right)^c = X \setminus \overline{V}$. So V and $\left(\overline{V}\right)^c = W$ are open sets satisfying $x \in V$, $A \subseteq W$ and $V \cap W = V \cap \left(\overline{V}\right)^c \subseteq V \cap V^c = \phi$. ($V \subseteq \overline{V}$ implies $\left(\overline{V}\right)^c \subseteq V^c$.) Hence by definition $(X, \mathcal{J})$ is a regular space.

In a similar way we prove the following theorem.

Theorem 9. A T_1-topological space is a normal space if and only if whenever A is a closed subset of X and U is an open set containing A, then there exists an open set V containing A such that $\overline{V} \subseteq U$.

Proof: Assume that $(X, \mathcal{J})$ is a normal topological space. Now take a closed set A and an open set U in X such that $A \subseteq U$. Now $A \subseteq U$ implies $U^c \subseteq A^c$. Here A, $U^c = B$ are closed sets such that $A \cap B = A \cap U^c \subseteq U \cap U^c = \phi$. That is A, B are disjoint closed subsets of the normal space $(X, \mathcal{J})$. Hence there exist open sets U, W in X such that $A \subseteq V$, $B = U^c \subseteq W$ and $V \cap W = \phi$. Further $\overline{V} \subseteq W^c$ (note: $V \subseteq W^c$ implies $\overline{V} \subseteq \overline{W^c} = W^c$). Now $\overline{V} \subseteq W^c \subseteq U$. Hence whenever A is a closed set and U is an open set containing A then there exists an open set V such that $A \subseteq V$, $\overline{V} \subseteq U$. Now let us assume that the above statement is satisfied. So our aim is to prove that $(X, \mathcal{J})$ is a normal space. So start with disjoint closed subsets say A, B of X. Now $A \cap B = \phi$ implies $A \subseteq B^c = U$. That is U is an open set containing the closed set A. Hence by our assumption there exists an open set V such that $A \subseteq V$, $\overline{V} \subseteq U$. Now $\overline{V} \subseteq U$ implies $U^c \subseteq \left(\overline{V}\right)^c$ implies $B \subseteq \left(\overline{V}\right)^c$. Further $V \cap \left(\overline{V}\right)^c \subseteq V \cap V^c = \phi$. That is whenever A, B are closed subsets of X, then there exist open sets V and $\left(\overline{V}\right)^c = W$ such that $A \subseteq V$, $B \subseteq W$ and $V \cap W = \phi$. Therefore by definition $(X, \mathcal{J})$ is a normal space.

Example of a Topological Space which is Regular but not Normal

Let $\mathcal{J}_l$ be a lower limit topology on $\mathbb{R}$. That is $\mathbb{R}_l = \left(\mathbb{R}, \mathcal{J}_l\right)$. Now let us prove that the product space $\mathbb{R}_l \times \mathbb{R}_l$ is a regular space. (If X, Y are regular topological spaces then the product space $X \times Y$ is a regular space. Hence it is enough to prove that $\mathbb{R}_l$ is a regular space.) For $(x, y) \in \mathbb{R}^2$, each basic open set U of the form $U = [x, a) \times [y, b)$ is both open and closed. Hence for each basic neighbourhood U of (x, y) in $\mathbb{R}_l \times \mathbb{R}_l$ there exists a neighbourhood $V = U$ of (x, y) such that $\overline{V} = \overline{U} \subseteq U$. Now if U' is any open set containing (x, y) then there exists a basic open set U as given above such that $(x, y) \in U = [x, a) \times [y, b) \subseteq U'$. Therefore $V = U$ is an open set containing (x, y) and $\bar{V} = \bar{U} = U \subseteq U'$. Also $\mathbb{R}_l \times \mathbb{R}_l$ is a Hausdorff space. Hence $\mathbb{R}_l \times \mathbb{R}_l$ is a regular space. Now let us take $Y = \{(x, y) \in \mathbb{R}^2 : y = -x\}$ then for each $(x, y) \in Y$ there exists $a, b \in \mathbb{R}$, $x < a$, $y < b$ such that $([x, a) \times [y, b)) \cap Y = \{(x, y)\}$. Hence each singleton $\{(x, y)\}$ is open in the subspace Y of $\mathbb{R}_l \times \mathbb{R}_l$. This proves that the subspace Y of $\mathbb{R}_l \times \mathbb{R}_l$ is discrete. Also Y is a closed subset of $\mathbb{R}_l \times \mathbb{R}_l$. Let $A = \{(x, y) \in \mathbb{R}^2 : y = -x \in \mathbb{Q}\}, B = \{(x, y) \in \mathbb{R}^2 : y = -x \in \mathbb{Q}^c\}$. Now A, B are closed sets in Y and Y is a closed set in $\mathbb{R}_l \times \mathbb{R}_l$ implies A, B are closed in $\mathbb{R}_l \times \mathbb{R}_l$. Also $A \cap B = \phi$. Suppose there exist open sets U, V in $\mathbb{R}_l \times \mathbb{R}_l$ satisfying $A \subseteq U$, $B \subseteq V$. Then we can observe that $U \cap V \neq \phi$. Therefore the product space $\mathbb{R}_l \times \mathbb{R}_l$ is not a normal space.

Remark: We can prove that $(\mathbb{R}, \mathcal{J}_l) = \mathbb{R}_l$ is a normal space. So, $\mathbb{R}_l \times \mathbb{R}_l$ is a regular space but it is not a normal space.

We have already proved that every compact subset of Hausdorff topological space is closed. Essentially we use the same proof technique used there to prove the following theorem:

Theorem 10. Every compact Hausdorff topological space $(X, \mathcal{J})$ is a regular space.

Proof: Let A be a closed subset of X and $x \in X \setminus A$, then for each $y \in A$, $x \neq y$. Hence X is a Hausdorff space implies that there exist open sets U_y, V_y in X satisfying $x \in U_y$, $y \in V_y$ and $U_y \cap V_y = \phi$. We know that closed subset of a compact space is compact. Here $A \subseteq \bigcup_{y \in A} V_y$.

That is $\{V_y : y \in A\}$ is an open cover for the compact space A. Therefore there exists $n \in \mathbb{N}$ and $y_1, y_2, \ldots, y_n \in A$ such that $A \subseteq \bigcup_{i=1}^{n} V_{y_i}$. Let $U = \bigcap_{i=1}^{n} U_{y_i}$ and $V = \bigcup_{i=1}^{n} V_{y_i}$. Then U, V are open sets in X satisfying $x \in U$, $A \subseteq V$ and $U \cap V \subseteq U \cap (V_{y_1} \cup V_{y_2} \cup \cdots \cup V_{y_n})$

$$= (U \cap V_{y_1}) \cup (U \cap V_{y_2}) \cup \cdots \cup (U \cap V_{y_n}) \subseteq (U_{y_1} \cap V_{y_1}) \cup (U_{y_2} \cap V_{y_2}) \cup \cdots \cup (U_{y_n} \cap V_{y_n}) = \phi .$$

Hence by definition $(X, \mathcal{J})$ is a regular space.

Now let us prove that every compact Hausdorff space is a normal space.

Theorem 11. Every compact Hausdorff space $(X, \mathcal{J})$ is a normal space.

Proof: Let A, B be disjoint closed sets in X. Then for each $x \in A$, $x \notin B$. Now $(X, \mathcal{J})$ is a regular space implies there exist open sets U_x, V_x satisfying: $x \in U_x$; $B \subseteq V_x$ and $U_x \cap V_x = \phi$.

Now $\{U_x : x \in A\}$ is an open cover for A implies there exists $n \in \mathbb{N}$, $x_1, x_2, \ldots, x_n \in A$ such that $A \subseteq \bigcup_{i=1}^{n} U_{x_i}$. Let $U = U_{x_1} \cup U_{x_2} \cup \cdots \cup U_{x_n}$ and $V = V_{x_1} \cup V_{x_2} \cup \cdots \cup V_{x_n}$. Then U, V are open sets in X satisfying $A \subseteq U$, $B \subseteq V$ and $U \cap V = \phi$. Hence by definition $(X, \mathcal{J})$ is a normal space.

Theorem 12. Closed subspace of a normal topological space $(X, \mathcal{J})$ is normal.

Proof: Let Y be a closed subspace of $(X, \mathcal{J})$. That is Y is a closed subset of $(X, \mathcal{J})$ and $\mathcal{J}_Y = \{A \cap Y : A \in \mathcal{J}\}$ is a topology on Y. So we will have to prove that $(Y, \mathcal{J}_Y)$ is a normal space. To prove this, take a closed set $A \subseteq Y$ and an open set U in $(Y, \mathcal{J}_Y)$ such that $A \subseteq U$. Now U is an open set in $(Y, \mathcal{J}_Y)$ implies there exists $V \in \mathcal{J}$ such that $U = V \cap Y$. Also A is a closed set in the subspace implies $A = \overline{A}_Y = \overline{A} \cap Y$ (here $\overline{A}_Y$ denotes the closure of A in $(Y, \mathcal{J}_Y)$ and $\overline{A}$ denotes the closure of A in $(X, \mathcal{J})$). Now $\overline{A}$, Y are closed sets in X implies $\overline{A} \cap Y$ is also a closed set in X. Hence A is a closed set in $(X, \mathcal{J})$ and V is an open set in $(X, \mathcal{J})$ containing A and $(X, \mathcal{J})$ is a normal topological space implies there exists an open set W in $(X, \mathcal{J})$ such that $A \subseteq W$ and $\overline{W} \subseteq V$. Now $W \cap Y$ is an open set in $(Y, \mathcal{J}_Y)$ and $A \subseteq W \cap Y$ and $\overline{W \cap Y} \subseteq \overline{W} \cap \overline{Y} \subseteq V \cap Y \subseteq U$. We started with a closed set A in $(Y, \mathcal{J}_Y)$ and an open set U in $(Y, \mathcal{J}_Y)$ such that $A \subseteq U$. Now we have proved that there exists an open set $W \cap Y$ in $(Y, \mathcal{J}_Y)$ satisfying $A \subseteq W \cap Y$ and $\left(\overline{W \cap Y}\right)_Y = \overline{W \cap Y} \cap Y = \overline{W \cap Y} \subseteq U$. That is $W \cap Y$ is an open set in the subspace containing A and closure of this open set with respect to the subspace $(Y, \mathcal{J}_Y)$ is contained in U. Hence $(Y, \mathcal{J}_Y)$ is a normal space.

Urysohn Lemma

Now let us prove the following important theorem known as Urysohn lemma.

Theorem 13. Let $(X, \mathcal{J})$ be a normal space and A, B be disjoint nonempty closed subsets of X. Then there exists a continuous function $f : X \to [0, 1]$ such that $f(x) = 0$ for every x in A, and $f(x) = 1$ for every x in B.

Proof: $A \cap B = \phi$ implies $A \subseteq B^c = X \backslash B$. Hence B^c is an open set containing the closed set A. Now X is a normal space implies there exists an open set U_0 such that $A \subseteq U_0$ and $\overline{U}_0 \subseteq B^c = U_1$. Now $[0, 1] \cap \mathbb{Q}$ is a countable set implies there exists a bijective function say $f : \mathbb{N} \to [0,1] \cap \mathbb{Q}$ satisfying $f(1) = 1$, $f(2) = 0$ and $f(\mathbb{N} \setminus \{1,2\}) = (0,1) \cap \mathbb{Q}$. That is $[0, 1] \cap \mathbb{Q} = \{r_1, r_2, r_3, \ldots\}$ such that $r_1 = 1$, $r_2 = 0$ and $f(k) = r_k$ for $k \geq 3$.

Aim: To define a collection $\{U_p\}_{p \in [0,1] \cap \mathbb{Q}}$ of open sets such that for $p, q \in [0,1] \cap \mathbb{Q}$, $p < q$ implies $\overline{U}_p \subseteq U_q$.

Let $P_n = \{r_1, r_2, \ldots, r_n\}$. Assume that U_p is defined for all $p \in P_n$, where $n \geq 2$ and this collection satisfies the property namely $p, q \in [0, 1] \cap \mathbb{Q}$, $p < q$ implies $\overline{U}_p \subseteq U_q$. Note that this result is true when $n = 2$. Now let us prove this result for P_{n+1}. Here $P_{n+1} = P_n \cup \{r_{n+1}\}$.

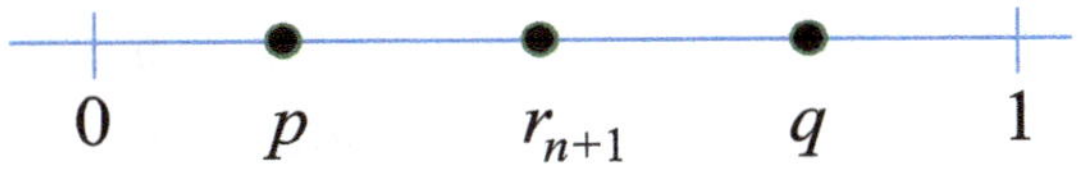

Let p, q ∈ P_{n+1} be such that p = max{r ∈ P_{n+1} : r < r_{n+1}} and q = min{r ∈ P_{n+1} : r > r_{n+1}}. Now $p,\ q \neq r_{n+1}$ implies p, q ∈ P_n. By our assumption U_p, U_q are known and $\bar{U}_p \subseteq U_q$. Now U_q is an open set containing the closed set $\bar{U}_p$ and X is a normal space. Hence there exists an open set say $U_{r_{n+1}}$ such that $\bar{U}_p \subseteq U_{r_{n+1}}$ and $\bar{U}_{r_{n+1}} \subseteq U_q$. If r, s ∈ P_n then we are through.

Suppose r ∈ P_n and s = r_{n+1} then r ≤ p or r ≥ q. If $r \leq p,\ \bar{U}_r \subseteq U_p \subseteq \bar{U}_p \subseteq U_s$. If $r \geq p,\ \bar{U}_s \subseteq U_q \subseteq \bar{U}_q \subseteq U_r$ and therefore by induction U_p is defined for all $p \in [0,\ 1] \cap \mathbb{Q}$ and $p, q \in [o,1] \cap \mathbb{Q}$, p < q implies $\bar{U}_p \subseteq U_q$.

Now define $U_p = \phi$, if $p \in \mathbb{Q},\ p < 0$ and U_p = X if $p \in \mathbb{Q},\ p > 1$. Then $p,\ q \in \mathbb{Q},\ p < q$ implies $\bar{U}_p \subseteq U_q$. Define f : X → [0, 1] as $f(x) = inf\{p \in \mathbb{Q} : x \in U_p\}$. Now x ∈ A, then x ∈ U_0. Hence x ∈ U_p for all p ≥ 0. In this case $\{p \in \mathbb{Q} : x \in U_p\} = [0, \infty) \cap \mathbb{Q}$. Hence inf $\{p \in \mathbb{Q} : x \in U_p\} = 0$.

That is x ∈ A implies $f(x)$ = 0. Now suppose $x \in B = U_1^c$ then $x \notin U_p$ for all p ≤ 1. Hence $\{p \in \mathbb{Q} : x \in U_p\} = [1,\ \infty) \cap \mathbb{Q}$ *implies* $f(x) = 1$ for all x ∈ B.

Now let us prove that f is a continuous function. $\mathcal{S}$ = {[0, a),(a, 1] : 0 < a < 1} is a subbase for [0, 1]. Hence it is enough to prove that for each a, 0 < a < 1, f^{-1}([0, a)) and f^{-1}((a, 1])) are open sets in X. For 0 < a < 1, let us prove that $f^{-1}([0,\ a)) = \{x \in X : 0 \leq f(x) < a\} = \bigcup_{p<a} U_p$. Now x ∈ f^{-1}([0, a)) implies $f(x)$ < a implies there exists a rational number p such that $f(x)$ < p < a. By the definition of $f(x)$, x ∈ U_p.

Hence

$$f^{-1}([0,\ a)) \subseteq \bigcup_{p<a} U_p. \tag{2}$$

Now let x ∈ U_p for p < a implies $f(x)$ ≤ p implies x ∈ f^{-1} ([0, a)). Hence we have

$$\bigcup_{p<a} U_p \subseteq f^{-1}([0,\ a)). \tag{3}$$

From Eqs. (2) and (3) we have $f^{-1}([0,\ a)) = \bigcup_{p<a} U_p$. Now each U_p is an open set implies that $\bigcup_{p<a} U_p$ is an open set in X. In a similar way we can prove that f^{-1}((a, 1]) is also an open subset of X for each 0 < a < 1. Now $f : X \to$ [0, 1] such that inverse image of each subbasic open set is an open set implies that $f : X \to$ [0, 1] is a continuous function.

Theorem 14. Let $(X, \mathcal{J})$ be a normal space and A, B be disjoint nonempty closed subsets of X. Then for $a,\ b \in \mathbb{R},\ a < b$ there exists a continuous function $f : X \to$ [a, b] such that $f(x) = a$ for every x in A, and $f(x) = b$ for every x in B.

Proof: Define g : [0, 1] → [a, b] as g(t) = a + (b − a)t then g is continuous. Now by theorem 13 there is a continuous function $f_1 : X \to [0, 1]$ such that $f_1(x) = 0$, for all $x \in A$ and $f_1(x) = 1$ for all $x \in B$. The function $f = g \circ f_1 : X \to [a, b]$ is a continuous function and further $f(x) = g(f_1(x)) = g(0) = a$ for all $x \in A$ and $f(x) = g(f_1(x)) = g(1) = b$ for all $x \in B$.

Remark: Let A, B be nonempty disjoint closed subsets of a metric space (X, d). Define $f : X \to \mathbb{R}\ as\ f(x) = \dfrac{d(x, A)}{d(x, A) + d(x, B)}$. Observe that f is a continuous function satisfying the condition that $f(x) = 0$ for all $x \in A$ and $f(x) = 1$ for all $x \in B$. It shows that the proof of Urysohn lemma is trivial (or say simple) if our topological space is a metrizable topological space.

Tietze Extension Theorem

Theorem 15. Tietze Extension Theorem. Let A be a nonempty closed subset of a normal space X and let f: A → [−1, 1] be a continuous function. Then there exists a continuous function g : X → [−1, 1] such that $g(x) = f(x)$ for all x in A.

Proof: The sets $\left[-1, \frac{-1}{3}\right], \left[\frac{1}{3}, 1\right]$ are closed subsets of [−1, 1] and f : A → [−1, 1] is a continuous function implies $A_1 = f^{-1}\left(\left[\frac{1}{3}, 1\right]\right), B_1 = f^{-1}\left(\left[-1, \frac{-1}{3}\right]\right)$ are closed subsets of the subspace A. (Here consider A as a subspace of X.) Now $x \in A_1 \cap B_1$ implies $f(x) \in \left[-1, \frac{-1}{3}\right] \cap \left[\frac{1}{3}, 1\right]$ a contradiction. Hence $A_1 \cap B_1 = \phi$. Now A_1, B_1 are closed in A and A is closed in X implies A_1, B_1 are closed in the normal space X. Hence by Urysohn's lemma there exists a continuous function $f_1 : X \to \left[\frac{-1}{3}, \frac{1}{3}\right]$ such that $f_1(A_1) = \frac{1}{3}$ and $f_1(B_1) = -\frac{1}{3}$ then $|f(x) - f_1(x)| \le \frac{2}{3}$ for all $x \in A$. Now consider the function $f - f_1 : A \to \left[\frac{-2}{3}, \frac{2}{3}\right]$ then $A_2 = (f - f_1)^{-1}\left(\left[\frac{2}{9}, \frac{2}{3}\right]\right)$ and $B_2 = (f - f)^{-1}\left(\left[\frac{-2}{3}, \frac{-2}{9}\right]\right)$ are disjoint closed subsets of X. By Urysohn lemma there exists a continuous function $f_2 : X \to \left[-\frac{2}{9}, \frac{2}{9}\right]$ such that $f_2(A_2) = \frac{2}{9}$ and $f_2(B_2) = -\frac{2}{9}$. Also

$\left|f(x)-\left(f_1(x)+f_2(x)\right)\right| \leq \frac{4}{9}$ for all $x \in$ A. By proceeding as above by induction for each $n \in \mathbb{N}$ there exists a continuous function $f_n : X \rightarrow \left[\frac{-2^{n-1}}{3^n}, \frac{2^{n-1}}{3^n}\right]$ such that

$$\left|f(x)-\sum_{i=1}^{n} f_i(x)\right| \leq \left(\frac{2}{3}\right)^n \text{ for all } x \in A. \qquad (4)$$

That is $f_n : \mathrm{X} \rightarrow [-1, 1]$ is a sequence of continuous functions such that $\left|f_n(x)\right| \leq \frac{2^{n-1}}{3^n} = M_n$ and $\sum^{\infty} M_n < \infty$ By Weierstrass M-test, the series $\sum_{n=1}^{\infty} f_n(x)$ converges uniformly on X. That is, if $s_n(x) = \sum_{i=1}^{n} f_i(x),\ x \in X$ then $s_n(x)$ converges uniformly on X. Also each $s_n : X \rightarrow \mathbb{R}$ is continuous. We know, from analysis, if a sequence $s_n : X \rightarrow \mathbb{R}$ of continuous functions converges uniformly to a function $g : X \rightarrow \mathbb{R}$ then g is also a continuous function. Hence $g : X \rightarrow \mathbb{R}$ be defined as $g(x) = \sum_{n=1}^{\infty} f_n(x)$ is continuous. Now for each $x \in A, \left|f(x)-\sum_{i=1}^{n} f_i(x)\right| \leq \left(\frac{2}{3}\right)^n$ (from (4)). Therefore $\left|g(x)-f(x)\right| = \left|\lim_{n\to\infty}\sum_{i=1}^{n} f_i(x)-f(x)\right| = \lim_{n\to\infty}\left|\sum_{i=1}^{n} f_i(x)-f(x)\right| \leq \lim_{n\to\infty}\left(\frac{2}{3}\right)^n = 0$. This implies $g(x) = f(x)$ for all $x \in$ A.

Definition: A topological space $(X, \mathcal{J})$ is said to be completely regular if (i) for each $x \in X$, singleton $\{x\}$ is closed in $(X, \mathcal{J})$ (that is $(X, \mathcal{J})$ is a T_1-space), (ii) for $x \in X$ and any nonempty closed set A with $x \notin A$ there exists a continuous function $f : X \rightarrow [0, 1]$ such that $f(x) = 0$ and $f(y) = 1$ for all $y \in A$.

Result: Every normal space $(X, \mathcal{J})$ is completely regular.

Proof: Let $x \in X$ and A be a nonempty closed set with $x \notin A$. Now $\{x\}$, A are disjoint closed sets. Hence by Urysohn's lemma there exists a continuous function $f : X \rightarrow [0, 1]$ such that $f(x) = 0$ and $f(y) = 1$ for all $y \in A$.

Result: If Y is a subspace of a completely regular space $(X, \mathcal{J})$ then $(Y, \mathcal{J}_Y)$ is also a completely regular space.

Proof: Let $y \in Y$ and A be a closed set in $(Y, \mathcal{J}_Y)$ with $y \notin A$. Since A is a closed set in Y there exists a closed set F in $(X, \mathcal{J})$ such that $A = F \cap Y$, $y \notin F$, F is a closed set in the completely regular space $(X, \mathcal{J})$ implies there exists a continuous function $f : X \rightarrow [0, 1]$ such that $f(y) = 0$ and $f(a) = 1$ for all $a \in F$. Now $f : X \rightarrow [0, 1]$ is a continuous function implies $f \mid Y = g : (Y, \mathcal{J}_Y) \rightarrow [0, 1]$ (here $g(x) = (f|Y)(x) = f(x)$ for all $x \in Y$) is a continuous function. Now $g : (Y, \mathcal{J}_Y) \rightarrow [0, 1]$ is a continuous function such that g(y) = f(y) = 0 and g(a) = f(a) = 1 for all $a \in A = F \cap Y$. Also subspace of a T_1-space (do it as an exercise) is T_1-space. Hence the subspace $(Y, \mathcal{J}_Y)$ is a completely regular space.

Baire Category Theorem

Baire category theorem has many applications in topology and analysis. Our aim here is to state and prove this theorem for locally compact Hausdorff topological spaces.

Definition: A subset A of a topological space $(X, \mathcal{J})$ is said to be nowhere dense in X if and only if $(\overline{A})^{\circ} = int(\overline{A}) = \phi$.

Example: (i) $\mathbb{N}$ is nowhere dense in $\mathbb{R}$ ($\mathbb{R}$ with standard topology).

(ii) $\mathbb{Q}$ is dense in $\mathbb{R}$, that is $\overline{\mathbb{Q}} = \mathbb{R}$, and hence $\mathbb{Q}$ is not nowhere dense in $\mathbb{R}$. Here we have $(\overline{\mathbb{Q}})^{\circ} = \mathbb{R}^{\circ} = \mathbb{R} \neq \phi$.

Definition: A topological space $(X, \mathcal{J})$ is said to be of first category if and only if there exists a countable collection $\{E_n\}_{n=1}^{\infty}$ of subsets of X satisfying:

(i) for each $n \in \mathbb{N}$, $(\overline{E_n})^{\circ} = \overline{E_n}^{\circ} = \phi$, and

(ii) $X = \bigcup_{n=1}^{\infty} E_n$.

Definition: A nonempty subset Y of a topological space $(X, \mathcal{J})$ is said to be of first category in X if and only if there exists a countable collection $\{E_n\}_{n=1}^{\infty}$ of subsets of X satisfying:

(i) for each $n \in \mathbb{N}$, $(\overline{E}_n)^{\circ} = \phi$, and

(ii) $Y = \bigcup_{n=1}^{\infty} E_n$.

Remark: If Y is a nonempty subset of a topological space $(X, \mathcal{J})$ then $(Y, \mathcal{J}_Y)$ ($\mathcal{J}_Y = \{U \cap Y : U \in \mathcal{J}\}$) is also a topological space. It is possible that a subset Y of a topological space $(X, \mathcal{J})$ is of first category in $(X, \mathcal{J})$ but the subspace $(Y, \mathcal{J}_Y)$ is not of first category.

Example: Let $X = \mathbb{R}$ and $\mathcal{J}_s$ be the standard topology on $\mathbb{R}$. Then $Y = \mathbb{N}$, the set of all natural numbers, is of first category in $\mathbb{R}$, but the subspace $(\mathbb{N}, \mathcal{J}_{s/\mathbb{N}})$ is not of first category. For each $n \in \mathbb{N}$ let $E_n = \{n\}$. As E_n contains only one element namely n, $(\overline{E_n})^{\circ} = \{n\}^{\circ} = \phi$ in $\mathbb{R}$. Also $\mathbb{N} = \bigcup_{n=1}^{\infty}\{n\} = \{1, 2, \ldots\} = \bigcup_{n=1}^{\infty} E_n$. Hence $\mathbb{N}$ is of first category in $(\mathbb{R}, \mathcal{J}_s)$. But note that the subspace $(\mathbb{N}, \mathcal{J}_{s/Y})$ is the discrete topological space on $\mathbb{N}$. For $n \in \mathbb{N}$, $(n-1, n+1)$

is an open set in $\mathbb{R}$ and hence $(n-1,\ n+1)\cap \mathbb{N}=\{n\}$ is an open set in the subspace $(\mathbb{N},\ J_{s/\mathbb{N}})$. Now it is easy to see that there cannot exist any countable collection $\{A_n\}_{n=1}^{\infty}$ of subsets of $\mathbb{R}$ satisfying $\left\{\overline{A_n}\right\}^{\circ}=\phi$ and $\bigcup_{n=1}^{\infty} A_n=\mathbb{N}$. Note that for $A_n\subseteq \mathbb{N},\left(\overline{A_n}\right)^{\circ}=A_n$ with respect to $(\mathbb{N},\ J_{s/\mathbb{N}})$ and hence $(\mathbb{N},\ J_{s/\mathbb{N}})$ is not of first category.

Definition: If a topological space $(X,\ J)$ is not of first category then we say that the topological space $(X,\ J)$ is of second category.

Note. We have seen that $\mathbb{N}$ is of first category in $(\mathbb{R},\ J_s)$ but the topological space $(\mathbb{N},\ J_{s/Y})$ is of second category.

Now our main aim is to prove that every locally compact Hausdorff topological space is of second category.

Remark: First let us prove that every locally compact Hausdorff topological space $(X,\ J)$ is a regular space. So let us take a closed set A in $(X,\ J)$ and a point $x \in X\setminus A$.

We have seen that every compact Hausdorff space is a normal space and hence every compact Hausdorff space is a regular space. We know that the one point of compactification $(X^*,\ J^*)$ of $(X,\ J)$ is a compact Hausdorff space and $J^*|_X = J$. That is $(X,\ J)$ is a subspace of the compact Hausdorff space $(X^*,\ J^*)$. Also it is easy to prove that subspace of a regular space is regular (and it is to be noted that subspace of a normal space need not be a normal space) and hence $(X^*,\ J^*)$ is a regular space implies the subspace $(X,\ J)$ of $(X^*,\ J^*)$ is also a regular space.

Now we are in a position to state and prove the main theorem.

Theorem 16. Baire Category Theorem. Let $(X,\ J)$ be a locally compact Hausdorff topological space and $\{E_n\}_{n=1}^{\infty}$ be a countable collection of open sets in $(X,\ J)$. Further suppose for each $n\in\mathbb{N}$, $\overline{E_n}=X$ (E_n is dense in X) then $\bigcap_{n=1}^{\infty} E_n$ is also dense in X. That is $\overline{\left(\bigcap_{n=1}^{\infty} E_n\right)}=X$.

Proof: We want to prove that $\bigcap_{n=1}^{\infty} E_n$ is dense in X.

So take $x \in X$ and an open set U containing x. Now $(X,\ J)$ is a locally compact Hausdorff space implies there exists an open set V containing x such that $\overline{V}$ is compact. Now let $U_0 = U \cap V$. Then U_0 is an open set containing x. Also $\overline{U_0}\subseteq\overline{V}$ implies $\overline{U_0}$ is a compact set (since closed subset of compact set is compact). Now our aim is to prove that $U\cap\left(\bigcap_{n=1}^{\infty} E_n\right)\neq\phi$. For each n, E_n is open and $\overline{E_n}=X$, that is each E_n is open and dense in X. Start with n = 1, now $x \in X = \overline{E_1}$ and U_0 is an open set containing x implies $U_0 \cap E_1 \neq \phi$. So take an element say $x_1 \in U_0 \cap E_1$. Now U_0, E_1 are open sets implies $U_0 \cap E_1$ is also an open set. Now $U_0 \cap E_1$ is an open set containing x_1 and $(X, \mathcal{J})$ is a regular space (every locally compact Hausdorff space is a regular space) implies there exists an open set U_1 in X satisfying $x_1 \in U_1$, $\overline{U_1} \subseteq U_0 \cap E_1$. Now $x_1 \in \overline{E_2} = X$ implies $U_1 \cap E_2 \neq \phi$. Let $x_2 \in U_1 \cap E_2$. Since X is a regular space implies there exists an open set U_2 in X satisfying $x_2 \in U_2$, $\overline{U_2} \subseteq U_1 \cap E_2$. Again $x_2 \in \overline{E_3} = X$ and U_2 is an open set containing x_2 implies $U_2 \cap E_3 \neq \phi$. Let $x_3 \in U_2 \cap E_3$. Choose an open set U_3 such that $x_3 \in U_3$, $\overline{U_3} \subseteq U_2 \cap E_3$. Continuing in this way (that is using induction) we get a sequence $\{x_n\}_{n=1}^{\infty}$ in X and a sequence of open sets $\{U_n\}_{n=1}^{\infty}$ satisfying $x_n \in U_n$, $\overline{U_n} \subseteq U_{n-1} \cap E_n$ for all $n \in \mathbb{N}$. Note that $\overline{U}_n \subseteq U \cap \left(\bigcap_{k=1}^{n} E_k \right)$ for all $n \in \mathbb{N}$. Then $\{\overline{U}_k\}_{k=1}^{\infty}$ is a sequence of nonempty closed subsets of X and hence of the compact subspace $\overline{U}_0$. Further $\overline{U}_{k+1} \subseteq \overline{U}_k$ for any $k \in \mathbb{N}$ implies $\{\overline{U}_k\}_{k=1}^{\infty}$ has finite intersection property. That is $\{\overline{U}_k\}_{k=1}^{\infty}$ is a family of closed subsets of the compact topological space $\overline{U}_0$ and further $\{\overline{U}_k\}_{k=1}^{\infty}$ has finite intersection property. Therefore $\bigcap_{k=1}^{\infty} \overline{U}_k \neq \phi$. Let $a \in \bigcap_{k=1}^{\infty} \overline{U}_k$. Then $a \in \overline{U}_k$ for all $k \in \mathbb{N}$ and hence $a \in U$. Also $a \in E_n$ for all $n \in \mathbb{N}$. So $a \in \bigcap_{n=1}^{\infty} E_n$. Thus $a \in U \cap \left(\bigcap_{n=1}^{\infty} E_n \right)$. That is for each $x \in X$ and for each open set U containing x, $U \cap \left(\bigcap_{n=1}^{\infty} E_n \right) \neq \phi$. Hence $x \in \overline{\left(\bigcap_{n=1}^{\infty} E_n \right)}$. This gives that $X \subseteq \overline{\left(\bigcap_{n=1}^{\infty} E_n \right)}$ and hence $X = \overline{\left(\bigcap_{n=1}^{\infty} E_n \right)}$, that is $\bigcap_{n=1}^{\infty} E_n$ is dense in X.

Exercise: Prove that a subset E of a topological space $(X, \mathcal{J})$ is nowhere dense in X (that is $\left(\overline{E}\right)^{\circ} = \phi$ if and only if $\left(\overline{E}\right)^{c}$ is dense in X.

Remark: It is known that every complete metric space X is of second category. The notion of completeness cannot be defined in a topological space. So we give the following version of Baire Category theorem for a locally compact Hausdorff topological space.

Theorem 17. Every nonempty locally compact Hausdorff topological space $(X, \mathcal{J})$ is of second category.

Proof: Proof by contradiction.

Suppose $(X, \mathcal{J})$ is of first category. Then there exists a countable collection $\{E_n\}_{n=1}^{\infty}$ of subsets of X satisfying $\overline{E_n}^{\circ} = \phi$ and $X = \bigcup_{n=1}^{\infty} E_n = \bigcup_{n=1}^{\infty} \overline{E}_n$. Therefore $X^c = \phi = \bigcap_{n=1}^{\infty} \overline{E}_n^c$ and hence $\phi = \overline{\phi} = \overline{\bigcap_{n=1}^{\infty} \overline{E}_n^c}$. But $\{\overline{E}_n^c\}_{n=1}^{\infty}$ is a countable collection of dense open sets, implies by theorem 15 $\overline{\bigcap_{n=1}^{\infty} \overline{E}_n^c} = X \neq \phi$. This contradicts $\phi = \overline{\phi} = \overline{\bigcap_{n=1}^{\infty} \overline{E}_n^c}$. Hence $(X, \mathcal{J})$ is of second category.

Now we are in a position to prove Urysohn metrization theorem that gives sufficient conditions under which a topological space is metrizable. Also it is interesting to note that the well known Nagata-Smirnov metrization theorem gives a set of necessary and sufficient conditions for metrizability of a topological space.

Urysohn Metrization Theorem

Theorem 18. Urysohn Metrization Theorem. Every normal space $(X, \mathcal{J})$ with a countable basis is metrizable.

Proof: Let $\mathcal{B} = \{B_1, B_2, \ldots, \}$ be a countable basis for $(X, \mathcal{J})$. Suppose $n, m \in \mathbb{N}$ are such that $\overline{B}_n \subseteq B_m$ then $\overline{B_n} \cap B_m^c = \phi$. Hence by Urysohn's lemma there exists a continuous function say $g_{n,m} : X \to \mathbb{R}$ such that

$$g_{n,m}(x) = 0 \quad \textit{for all } x \in B_m^c, \tag{5}$$

and

$$g_{n,m}(x) = 1 \quad \textit{for all } x \in \overline{B}_n, \tag{6}$$

Now take $x_0 \in X$ and an open set U containing x_0. Since $\mathcal{B}$ is a basis for $(X, \mathcal{J})$ there exists $B_m \in \mathcal{B}$ such that $x_0 \in B_m \subseteq U$. Now B_m is an open set containing x_0 implies there exists an open set V containing x_0 such that $\overline{V} \subseteq B_m$. Hence there exists a basic open set B_n containing x_0 such that $\overline{B}_n \subseteq \overline{V} \subseteq B_m$. Hence for such pair (n, m) we have a continuous function $g_{n,m} : X \to \mathbb{R}$ satisfying Eq. (5).

So if $x_0 \in X$ and U is an open set containing x_0 then there exists a continuous function $g_{n,m} : X \to \mathbb{R}$ such that $g_{n,m}(x_0) = 1$ and $g_{n,m}(x) = 0$ for all $x \in U^c \subseteq B_m^c$. So we have proved that there exists a countable collection of continuous functions $f_n : X \to [0, 1]$ such that for $x_0 \in X$ and open set U containing x_0, there exists $n \in \mathbb{N}$ such that $f_n(x_0) = 1 > 0$ and $f_n(x) = 0$ for all $x \in U^c$. It is to be noted that $\{(n, m) : n, m \in \mathbb{N}\}$ is a countable set. We know that $\mathbb{R}^w = \mathbb{R} \times \mathbb{R} \times \mathbb{R} \times \cdots$ with product topology is metrizable. That is there is a metric d on $\mathbb{R}^w$ such that $\mathcal{J}_d$, the topology on $\mathbb{R}^w$ induced by d, coincides with the product topology on $\mathbb{R}^w$.

Now let us define a map $T : X \to \mathbb{R}^w$ as $T(x) = (f_1(x), f_2(x), \ldots,)$ and using this map we define $d_1(x, y) = d(T(x), T(y))$ and conclude that $J_{d_1} = J$. This will prove that (X, J) is a metrizable topological space. Now let us prove that (X, J) is homeomorphic to a subspace of $\mathbb{R}^w$. Each $f_n : X \to \mathbb{R}$ is a continuous function implies $T(x) = (f_1(x), f_2(x), \ldots,)$ is a continuous function.

To prove T is injective (one-one).

Let $x, y \in X$ be such that $x \neq y$. Then there exist open sets $U, V \in X$ such that $x \in U, y \in V$ and $U \cap V = \phi$. Now U is an open set containing x implies there exists $n \in \mathbb{N}$ such that $f_n(x) = 1$ and $f_n(y) = 0$ (note that $y \in U^c$). This implies $f_n(x) \neq f_n(y)$ for this particular $n \in \mathbb{N}$ and hence $(f_1(x), f_2(x), \ldots, f_n(x), \ldots) \neq (f_1(y), f_2(y), \ldots, f_n(y), \ldots)$. This means $Tx \neq Ty$. That is $x, y \in X$, $x \neq y$ implies $Tx \neq Ty$. This implies T is 1-1.

Now it is enough to prove that T maps open set A in X to an open set T(A) in Y = T(X). Let A be an open set and $y_0 \in T(A)$. Now $y_0 \in T(A)$ implies there exists $x_0 \in A$ such that $T(x_0) = y_0$. Now $x_0 \in A$, A is an open set implies there exists $n_0 \in \mathbb{N}$ such that $f_{n_0}(x_0) = 1$ and $f_{n_0}(x) = 0$ for all $x \in A^c$. We know that for each $n \in \mathbb{N}$ the projection map $p_n : \mathbb{R}^w \to \mathbb{R}$ defined as $p_n\left((x_k)_{k=1}^{\infty}\right) = x_n$ is a continuous map. Hence $(0, \infty)$ is an open set implies $V = p_{n_0}^{-1}((0, \infty))$ is an open subset of $\mathbb{R}^w$. This implies $V \cap Y$ is an open set in Y.

Now let us prove that $y_0 \in V \cap Y$ and $V \cap Y \subseteq T(A)$. $p_{n_0}(y_0) = (p_{n_0} \cdot T)(x_0) = f_{n_0}(x_0) = 1 > 0$ implies $y_0 \in V$. Also $y_0 \in Y$. Hence $y_0 \in V \cap Y$. That is $V \cap Y$ is an open set in Y containing the point y_0.

Now we claim that $V \cap Y \subseteq T(A)$. So, let $y \in V \cap Y$. Then there exists $x \in X$ such that $y = Tx$. This implies $p_{n_0}(y) \in (0, \infty)$ and $p_{n_0}(y) = p_{n_0}(T(x)) = f_{n_0}(x) \in (0, \infty)$. Hence $x \in A$ ($f_{n_0}(x) = 0$ *for* $x \in A^c$). So we have proved that $y = Tx \in V \cap Y$ implies $y = Tx \in T(A)$. Hence $V \cap Y$ is an open set in Y containing Tx and this set is contained in T(A). Therefore T(A) is open in Y. Hence we have proved that $T : (X, J) \xrightarrow{onto} (Y, d_Y)$ is a homeomorphism. (Here (Y, d_Y) is a subspace of $(\mathbb{R}^w, d)$.) Now $d_1(x, y) = d(Tx, Ty)$ for all $x, y \in X$ implies d_1 is a metric on X. Also it is easy to see that a subset A of X is open in (X, J) if and only if A is open in (X, J_{d_1}). Therefore $J_{d_1} = J$.

Permissions

All chapters in this book are published with permission under the Creative Commons Attribution Share Alike License or equivalent. Every chapter published in this book has been scrutinized by our experts. Their significance has been extensively debated. The topics covered herein carry significant information for a comprehensive understanding. They may even be implemented as practical applications or may be referred to as a beginning point for further studies.

We would like to thank the editorial team for lending their expertise to make the book truly unique. They have played a crucial role in the development of this book. Without their invaluable contributions this book wouldn't have been possible. They have made vital efforts to compile up to date information on the varied aspects of this subject to make this book a valuable addition to the collection of many professionals and students.

This book was conceptualized with the vision of imparting up-to-date and integrated information in this field. To ensure the same, a matchless editorial board was set up. Every individual on the board went through rigorous rounds of assessment to prove their worth. After which they invested a large part of their time researching and compiling the most relevant data for our readers.

The editorial board has been involved in producing this book since its inception. They have spent rigorous hours researching and exploring the diverse topics which have resulted in the successful publishing of this book. They have passed on their knowledge of decades through this book. To expedite this challenging task, the publisher supported the team at every step. A small team of assistant editors was also appointed to further simplify the editing procedure and attain best results for the readers.

Apart from the editorial board, the designing team has also invested a significant amount of their time in understanding the subject and creating the most relevant covers. They scrutinized every image to scout for the most suitable representation of the subject and create an appropriate cover for the book.

The publishing team has been an ardent support to the editorial, designing and production team. Their endless efforts to recruit the best for this project, has resulted in the accomplishment of this book. They are a veteran in the field of academics and their pool of knowledge is as vast as their experience in printing. Their expertise and guidance has proved useful at every step. Their uncompromising quality standards have made this book an exceptional effort. Their encouragement from time to time has been an inspiration for everyone.

The publisher and the editorial board hope that this book will prove to be a valuable piece of knowledge for students, practitioners and scholars across the globe.

Index

www.ingramcontent.com/pod-product-compliance
Lightning Source LLC
Chambersburg PA
CBHW061959190326
41458CB00009B/2913
9781635497045